CHEMISTRY REVIEW

IN 20 MINUTES A DAY

CHEMISTRY REVIEW IN 20 MINUTES A DAY

Second Edition

Published in the United States by LearningExpress, LLC, New York.

Library of Congress Cataloging-in-Publication Data

Chemistry review in 20 minutes a day.—2nd ed.
p. cm.
ISBN-13: 978-1-57685-799-1
ISBN-10: 1-57685-799-9
1. Chemistry—Study and teaching. I. LearningExpress (Organization)
II. Title: Chemistry review in twenty minutes a day.
QD4.0.C4536 2011
540—dc22

2010044945

Printed in the United States of America

9 8 7 6 5 4 3 2 1

Second Edition

ISBN 978-1-57685-799-1

For more information or to place an order, contact LearningExpress at:
2 Rector Street
26th Floor
New York, NY 10006

Or visit us at:
www.learnatest.com

CONTENTS

CONTENTS

INTRODUCTION

Welcome to the world of chemistry. Chemistry is around you all the time and impacts everything you do. Chemical reactions are helping you hold this book, understand the words on this page, and digest the meal you recently ate. *Chemistry Review in 20 Minutes a Day* is not intended to be a comprehensive look at chemistry or replace a comprehensive textbook in chemistry. It is anticipated that you use this book to help learn the basic concepts of chemistry that will enable you to better understand the comprehensive textbooks and more advanced materials. In addition, this book will give you the tools to recognize chemistry in everyday life. What is necessary before you start to learn and comprehend chemistry?

- **Appreciation of science/chemistry.** Even if you do not enjoy science, you need to have an appreciation for the need of science. Everyone's lives change daily with new research and development findings. The sender and receiver of news must fully understand science to have properly informed citizens.
- **Basic math and algebra skills.** Chemistry requires calculations and the manipulation of mathematical equations to solve problems. Review your algebra skills before starting the chemistry lessons and you will find that chemistry will be easier to comprehend.
- **No fear.** Chemophobia, or the fear of chemistry or chemicals, will not help you succeed in chemistry. Many people fear these terms because they are more commonly used to describe dangerous or toxic substances. In fact, everything is a chemical, including that water coming out of your sink. What, water is a chemical? Yep! Once again, everything is a chemical, and chemistry is the study to understand the substances that make up the world and universe.
- **Understand that chemistry is difficult, but not impossible.** Chemistry is an applied science and cannot be learned like most the subjects you learn in school. You also cannot "see" chemistry like you can biology, astronomy, or even physics. Electron microscopes can let a person see close to the atomic level, but you still cannot hold and view an atom. A little work and patience can allow you to see the patterns that

develop in chemistry and be able to apply those patterns to easy and more complex problems. You've encountered and conquered many difficult problems in life; chemistry is next!

How to Use This Book

Chemistry Review in 20 Minutes a Day teaches basic chemistry concepts in 24 self-paced lessons. The book includes a pretest, a posttest, tips on how to prepare for a standardized test, a glossary to help you recognize and remember the key chemistry concepts, and additional resources in case you would like to enhance your chemistry knowledge beyond the skills you will have learned in this book. Before you begin Lesson 1, take the pretest. The pretest will assess your current chemistry abilities. This will be helpful in determining your strengths and weaknesses. You'll find the answer key at the end of the pretest section.

After taking the pretest, move on to Lesson 1. Each lesson offers detailed explanations of new concepts. There are numerous examples with step-by-step solutions. As you proceed through a lesson, you will find tips and shortcuts that will help you learn a concept. Each new concept is followed by a practice set of problems. The answers to the practice problems are in an answer key located at the end of the book.

When you have completed all 24 lessons, take the posttest. The posttest has the same format as the pretest, but the questions are different. Compare the results of the posttest with the results of the pretest you took before you began Lesson 1. What are your strengths? Do you have weak areas? Do you need to spend more time on some concepts, or are you ready to go to the next level?

Make a Commitment

Success does not come without effort. If you truly want to be successful, make a commitment to spend the time you need to improve your chemistry skills.

So sharpen your pencils and get ready to begin the pretest!

PRETEST

Before you begin Lesson 1, "A Chemist's View of Chemistry," you may want to get an idea of what you know and what you need to learn. The pretest will answer some of these questions for you. The pretest consists of 30 multiple-choice questions covering the topics in this book. Although 30 questions can't cover every concept, skill, or shortcut taught in this book, your performance on the pretest will give you a good indication of your strengths and weaknesses. Keep in mind, the pretest does not test all the skills taught in this chemistry book.

If you score high on the pretest, you have a good foundation and should be able to work your way through the book quickly. If you score low on the pretest, don't despair. This book will take you through the chemistry concepts step by step. If you get a low score, you may need to take more than 20 minutes a day to work through a lesson. However, this is a self-paced program, so you can spend as much time on a lesson as you need. You decide when you fully comprehend the lesson and are ready to go on to the next one.

Take as much time as you need to do the pretest. When you are finished, check your answers with the answer key at the end of the pretest. You will find that the level of difficulty increases as you work your way through the pretest.

Answer Sheet

1.	a	b	c	d	**11.**	a	b	c	d	**21.**	a	b	c	d
2.	a	b	c	d	**12.**	a	b	c	d	**22.**	a	b	c	d
3.	a	b	c	d	**13.**	a	b	c	d	**23.**	a	b	c	d
4.	a	b	c	d	**14.**	a	b	c	d	**24.**	a	b	c	d
5.	a	b	c	d	**15.**	a	b	c	d	**25.**	a	b	c	d
6.	a	b	c	d	**16.**	a	b	c	d	**26.**	a	b	c	d
7.	a	b	c	d	**17.**	a	b	c	d	**27.**	a	b	c	d
8.	a	b	c	d	**18.**	a	b	c	d	**28.**	a	b	c	d
9.	a	b	c	d	**19.**	a	b	c	d	**29.**	a	b	c	d
10.	a	b	c	d	**20.**	a	b	c	d	**30.**	a	b	c	d

1. Which of the following statements about atoms is true?
   a. They have more protons than electrons.
   b. They have more electrons than protons.
   c. They are electrically neutral.
   d. They have as many neutrons as they have electrons.

2. What is the mass number of an atom with 60 protons, 60 electrons, and 75 neutrons?
   a. 120
   b. 135
   c. 75
   d. 195

3. According to Dalton's Theory, the only way a compound can consist of its elements in a definite ratio by mass is that it is made from the elements in
   a. a definite ratio by volume.
   b. a definite ratio by number of atoms.
   c. multiple whole-number ratios by mass.
   d. multiple whole-number ratios by volume.

4. Which of the following is a mixture?
   a. sodium chloride
   b. rice and beans
   c. magnesium sulfate
   d. water

5. Give the number of protons, neutrons, and electrons of this isotope of oxygen: $^{17}_{8}O$.
   a. 8 protons, 8 neutrons, 17 electrons
   b. 8 protons, 8 neutrons, 9 electrons
   c. 8 protons, 17 neutrons, 8 electrons
   d. 8 protons, 9 neutrons, 8 electrons

6. If the electron configuration of an element is written as $1s^2 2s^2 2p_x^2 2p_y^2 2p_z^2 3s^1$, the element's atomic
   a. number is 11.
   b. number is 12.
   c. weight is 11.
   d. weight is 12.

7. Choose the proper group of symbols for the following elements: potassium, silver, mercury, lead, sodium, iron.
   a. Po, Ar, Hr, Pm, So, Fm
   b. Pb, Sl, Me, Le, Su, Io
   c. Pt, Sr, My, Pd, Sd, In
   d. K, Ag, Hg, Pb, Na, Fe

8. What is the maximum number of electrons that the second energy level can hold?
   a. 8
   b. 6
   c. 2
   d. 16

9. The horizontal rows of the periodic table are called
   a. families.
   b. groups.
   c. representative elements.
   d. periods.

10. Which elements best conduct electricity?
    a. metals
    b. nonmetals
    c. metalloids
    d. ions

11. Knowing the *group* of an atom in the periodic table, how would you find the number of *valence electron(s)* for that atom?

a. The group number is equal to the number of valence electron(s) for any atom.

b. The group number is equal to the number of bond(s) an atom can form.

c. The group number indicates the number of orbitals for an atom.

d. The group number is equal to the number of shells in the atom.

12. The bond between oxygen and hydrogen atoms in a water molecule is a(n)

a. hydrogen bond.

b. polar covalent bond.

c. nonpolar covalent bond.

d. ionic bond.

13. Which of the following contains the formulas for these ions in this order: ammonium, silver, bicarbonate-/hydrogen carbonate, nitrate, calcium, fluoride?

a. Am^-, Si^{2+}, HCO_3^-, Na^+, CM^-, F^+

b. AM^+, Ag^+, CO_3^{2-}, NO_3^-, Cal^+, Fl^-

c. NH_4^-, Ag^+, HCO_3^-, NO_3^-, Cal^+, Fl^-

d. NH_4^+, Ag^+, HCO_3^-, NO_3^-, Ca^{2+}, F^-

14. How many electrons do the following have in their outer levels: S^{2-}, Na^+, Cl^-, Ar, Mg^{2+}, Al^{3+}?

a. three

b. five

c. seven

d. eight

15. What type of bond is formed when *electrons are shared* between two atoms?

a. shared bond

b. ionic bond

c. covalent bond

d. multiple bond

16. Which of the following molecules is nonpolar?

a. NH_3 and N_2

b. CO_2 and NO_2

c. NH_3 and NO_2

d. N_2 and CO_2

17. The molecular weight (in amu) of aluminum carbonate, $Al_2(CO_3)_3$, is

a. 55.

b. 114.

c. 234.

d. 201.

18. A sample of 11 g of CO_2 contains

a. 0.25 g of carbon.

b. 1.5 g of carbon.

c. 3.0 g of carbon.

d. 8.0 g of carbon.

19. How many grams are contained in 0.200 moles of calcium phosphate, $Ca_3(PO_4)_2$?

a. 6.20

b. 62.0

c. 124

d. 31.0

20. In the reaction, $CaCl_2 + Na_2CO_3 \rightarrow CaCO_3 + 2NaCl$, if you want to form 0.5 moles of NaCl, then

a. 1 mole of Na_2CO_3 is needed.

b. 0.5 moles of $CaCO_3$ are also formed.

c. 0.5 moles of Na_2CO_3 are needed.

d. 0.25 moles of $CaCl_2$ are needed.

21. In the reaction $2Cu_2S + 3O_2 \rightarrow 3Cu_2O + 2SO_2$, if 24 moles of Cu_2O are to be prepared, then how many moles of O_2 are needed?

a. 24
b. 36
c. 16
d. 27

22. Which of the following equations is balanced?

a. $2H_2O_2 \rightarrow 2H_2O + O_2$
b. $Ag + Cl_2 \rightarrow 2AgCl$
c. $KClO_3 \rightarrow KCl + O_2$
d. $Na + H_2O \rightarrow NaOH + H_2$

23. Butane (C_4H_{10}) burns with oxygen in air according to the following equation: $2C_4H_{10} + 13O_2 \rightarrow 8CO_2 + 10H_2O$. In one experiment, the supply of oxygen was limited to 98.0 g. How much butane can be burned by this much oxygen?

a. 15.1 g C_4H_{10}
b. 27.3 g C_4H_{10}
c. 54.6 g C_4H_{10}
d. 30.2 g C_4H_{10}

24. What type of chemical equation is $2NH_3 \rightarrow N_2 + 3H_2$?

a. combination reaction
b. decomposition reaction
c. single-displacement reaction
d. double-displacement reaction

25. A pressure of 740 mm Hg is the same as

a. 1 atm.
b. 0.974 atm.
c. 1.03 atm.
d. 0.740 atm.

26. What volume will 500 mL of gas—initially at 25° C and 750 mm Hg—occupy when conditions change to 25° C and 650 mm Hg?

a. 477 mL
b. 400 mL
c. 577 mL
d. 570 mL

27. Which law predicts that if the temperature (in Kelvin) doubles, the pressure will also double?

a. Boyle's Law
b. Charles's Law
c. Gay-Lussac's Law
d. Dalton's Law

28. Which of the following is NOT characteristic of gases?

a. They have a definite volume and shape.
b. They are low in density.
c. They are highly compressible.
d. They mix rapidly.

29. Gases that conform to the assumptions of kinetic theory are referred to as

a. kinetic gases.
b. natural gases.
c. ideal gases.
d. real gases.

30. A sample of helium at 25° C occupies a volume of 725 mL at 730 mm Hg. What volume will it occupy at 25° C and 760 mm Hg?

a. 755 mL
b. 760 mL
c. 696 mL
d. 730 mL

Answers

1. c.
2. b.
3. b.
4. b.
5. d.
6. a.
7. d.
8. a.
9. d.
10. a.
11. a.
12. b.
13. d.
14. d.
15. c.
16. d.
17. c.
18. c.
19. b.
20. d.
21. a.
22. a.
23. b.
24. b.
25. b.
26. c.
27. c.
28. a.
29. c.
30. c.

LESSON

A Chemist's View of Chemistry

Science attempts to understand and explain our observations in nature and the universe. Chemistry, a branch of science, studies the chemical and physical properties of matter. Chemistry plays a fundamental role in understanding particular concepts of biology, physics, astronomy, geology, and other disciplines of science. Because chemistry connects so many different fields, it is often referred to as the central science.

Chemistry—The Central Science

Chemistry is the study of the structure and properties of matter and the changes it undergoes during chemical reactions. These changes could be something as simple as dissolving sugar in water or as complex as photosynthesis. Chemistry provides the basic tools for scientists to be able to understand and interpret a vast array of physical processes.

A strong background in chemistry will help you to succeed not only in the classroom, but also in many careers—chemistry plays a huge role in biology, medicine, physics, energy, and a host of other fields. Chemistry is everywhere and, once you have completed this book, you will be able to see chemistry all around you in your everyday life.

The Scientific Method

Science is the search for new truths. To find this truth, scientists make observations, perform experiments, interpret results, and use their findings to develop new theories and laws. Chemists, like all scientists, follow a well-defined protocol, called the scientific method, to guide them through this process. The scientific method is divided into eight steps:

1. Identify a problem.
2. Research the problem.
3. Form a hypothesis.
4. Test the hypothesis by experiment
5. Collect and analyze the data.
6. Form a conclusion.
7. Publish the result.
8. Repeat the experiment.

By following this method, researchers help to ensure that conclusions drawn from experiments are interpreted without bias. While it may seem boring and redundant, the final step, repetition, is one of the most important. For an experiment's results to be accurate, it must be repeatable. How can a conclusion be true if the experiment that proves it yields a different result every time?

Practice

Identify which step of the scientific method corresponds to the following actions:

1. A man believes that white candles burn faster than brown candles.
2. A young girl wants to know if the amount of sunlight has an effect on how plants grow, so she goes to the library to find books on the subject.
3. A scientist reads about the results of an experiment in a scientific journal, so he decides to conduct the experiment on his own and check the results.
4. Based on the measurements of how long it takes for objects to fall from a known height, Isaac Newton concludes that the acceleration of a falling object due to gravity is 9.8 m/s^2.

Scientific Theories and Scientific Laws

When a hypothesis has been validated through numerous experiments and is generally accepted as accurate, the hypothesis can be classified as a *theory* or a *law*. The terms theory and law have particular, distinct meanings in science that differ from the common usage by the layperson.

A scientific law is a succinct mathematical or verbal description of an observed phenomenon. For example, Hooke's law, which describes the force required to stretch a spring follows the equation $\mathbf{F} = \mathbf{kx}$, where $\mathbf{F}$ is the applied force, $\mathbf{k}$ is the force constant of the spring, and $\mathbf{x}$ is the distance stretched. This equation is the result of numerous observations and has been shown to be consistent with many experiments.

A scientific theory takes the available evidence and observations and uses them to explain *why* something occurs. For example, the Big Bang Theory uses the results of many observations and experiments performed by astronomers and physicists to explain the origin of the universe.

Scientific theories, like scientific laws, are testable. That is, one can design experiments to verify that a law or theory is correct. In fact, the testing of theories and laws as new experimental techniques arise is an integral part of the scientific process. For example, the laws of classical mechanics held up for many years to describe the motion of particles until new technology allowed scientists to examine objects that are very small as well as objects that move very fast (close to the speed of light). These measurements established that classical mechanics does not hold up under these extreme circumstances, and led to the development of two new theories: quantum theory and the theory of relativity.

Practice

Identify the following as a *law* or a *theory*:

5. Mass is always conserved: The total mass of a chemical change remains the same.
6. Atoms and electrons possess particle-like and wave-like properties.
7. The attractive force due to gravity is $F = \frac{GMm}{r^2}$.
8. The volume of a gas is directly proportional to the number of particles.

Matter

Matter is anything that occupies space and has mass. In other words, matter is everything in the universe. Central to matter is the atom, which is the smallest unit of an element that retains its chemical properties. Atoms are made from subatomic particles (protons, neutrons, and electrons) and combine to form more complex molecules and compounds that make up all objects and substances that exist in nature.

Atom: The basic unit of an element that retains all the element's chemical properties. An atom is composed of a nucleus (which contains one or more protons and neutrons) and one or more electrons in motion around it. Atoms are electrically neutral because they are made of an equal number of protons and electrons.

Proton: A particle that has a mass of 1 atomic mass unit (amu; 1 amu = 1.66×10^{-27} kg) and an effective positive charge of +1.

Neutron: A particle that has a mass of 1 amu with no charge.

Electron: A particle that is of negligible mass (0.000549 amu) compared to the mass of the nucleus and that has an effective negative charge of -1.

Element: A substance that contains one type of atom and cannot be broken down by simple means.

Molecule: A combination of two or more atoms held together by a covalent bond. Molecules cannot be separated by physical means.

Compound: A combination of two or more atoms of different elements in a precise proportion by mass. A compound is held together by ionic bonds (ionic compound) or by covalent bonds (covalent compound). Atoms in a compound cannot be separated by physical means.

Matter can also be classified as one of four states: solid, liquid, gas, or plasma. To simplify, the discussion will be limited to solids, liquids, and gases (see Table 1.1). A *solid* is rigid and has a fixed volume. A *liquid* has a fixed volume but assumes the shape of its container. A *gas* has no definite shape or volume and can be compressed.

Table 1.1 States of Matter

STATE	VOLUME	SHAPE	COMPRESSIBILITY
Solid	Fixed	Rigid	No
Liquid	Fixed	Defined by container	No
Gas	Variable	Variable	Yes, highly

The Periodic Table of Elements

Approximately 115 elements have been discovered to date. These elements are organized in a periodic table of elements. Two seventeenth-century chemists, Dmitri Mendeleev and Julius Meyer, independently organized the early periodic table that evolved into the modern periodic table of elements (see Figure 1.2).

The periodic table of elements is structured according to the properties of the elements. Mendeleev's early experiments classified the elements according to their properties and reactivity with oxygen and grouped the elements in octaves (eight). Elements have chemical symbols that are used for their representation in the periodic table (O is oxygen and Zn is zinc). Some of the symbols are based on their original names, usually their Latin origin (Fe is iron [ferrum] and Na is sodium [natrium]).

The elements are organized by periods (horizontal rows) and *groups* (vertical columns). Elements in the same group (or family) usually have similar chemical properties, and they are identified by group numbers 1 through 18. The groups labeled 1A through 8A are often called *main group elements*. Elements in the same group also have (in common) the same number of electron(s) in their outermost (valence) shell (i.e., group 6A elements have six electrons in their outer shell). Groups 1A, 2A, 7A, and 8A have specific names based on their properties:

- Group 1A: Alkali metals (Li, Na, K, Rb, Cs, Fr)
- Group 2A: Alkaline earth metals (Be, Mg, Ca, Sr, Ba, Ra)
- Group 7A: Halogens (F, Cl, Br, I, At)
- Group 8A: Noble gases (He, Ne, Ar, Kr, Xe, Rn)

Elements in the same period have the same number of electron shells (or levels). Seven periods can be found in the modern periodic table.

Elements in the middle and left side of the table are classified as metals (Na, Fe, Hg). A *metal* is an element that is shiny, conducts electricity and heat, is malleable (easily shaped), and is ductile (pulled into wires). Metals are electropositive, having a greater tendency to lose their valence electrons.

Elements in the upper-right corner of the table are classified as nonmetals (C, F, S). A *nonmetal* is an element with poor conducting properties. Nonmetals are electronegative, having a greater tendency to gain valence electrons.

A *metalloid* or *semimetal* is an element with properties that are intermediate between those of metals and nonmetals, such as semiconductivity. They are found between metals and nonmetals in the periodic table (see shaded elements in Figure 1.2).

In addition to providing a convenient way to group elements together, the periodic table gives a great deal of information about each element. Each box within the table contains four pieces of information about a particular element: the atomic number, the elemental symbol, the element name, and the atomic mass. The atomic number tells the number of protons in the atom. The elemental symbol and element name provide two ways the element is identified. The atomic mass gives the average mass of the element (protons + neutrons). Fractional values arise due to an element having multiple isotopes, a topic discussed in the following lesson.

Key: 6 = Atomic number; C = Elemental symbol; Carbon = Element name; 12.011 = Atomic mass

1 1A	2 2A	3	4	5	6	7	8	9	10	11	12	13 3A	14 4A	15 5A	16 6A	17 7A	18 8A
1 H Hydrogen 1.008																	2 He Helium 4.003
3 Li Lithium 6.941	4 Be Beryllium 9.012											5 B Boron 10.811	6 C Carbon 12.011	7 N Nitrogen 14.007	8 O Oxygen 16.00	9 F Fluorine 19.00	10 Ne Neon 20.179
11 Na Sodium 22.99	12 Mg Magnesium 24.305											13 Al Aluminum 26.982	14 Si Silicon 28.086	15 P Phosphorus 30.974	16 S Sulfur 32.066	17 Cl Chlorine 35.45	18 Ar Argon 39.95
19 K Potassium 39.098	20 Ca Calcium 40.08	21 Sc Scandium 44.956	22 Ti Titanium 47.867	23 V Vanadium 50.942	24 Cr Chromium 51.996	25 Mn Manganese 54.938	26 Fe Iron 55.845	27 Co Cobalt 58.933	28 Ni Nickel 58.69	29 Cu Copper 63.546	30 Zn Zinc 65.39	31 Ga Gallium 69.72	32 Ge Germanium 72.61	33 As Arsenic 74.922	34 Se Selenium 78.96	35 Br Bromine 79.904	36 Kr Krypton 83.80
37 Rb Rubidium 85.468	38 Sr Strontium 87.62	39 Y Yttrium 88.906	40 Zr Zirconium 91.22	41 Nb Niobium 92.906	42 Mo Molybdenum 95.94	43 Tc Technetium 98.906	44 Ru Ruthenium 101.07	45 Rh Rhodium 102.906	46 Pd Palladium 106.42	47 Ag Silver 107.868	48 Cd Cadmium 112.41	49 In Indium 114.82	50 Sn Tin 118.71	51 Sb Antimony 121.76	52 Te Tellurium 127.60	53 I Iodine 126.905	54 Xe Xenon 131.30
55 Cs Cesium 132.906	56 Ba Barium 137.34	57 La * Lanthanum 138.906	72 Hf Hafnium 178.49	73 Ta Tantalum 180.95	74 W Tungsten 183.84	75 Re Rhenium 186.2	76 Os Osmium 190.23	77 Ir Iridium 192.22	78 Pt Platinum 195.078	79 Au Gold 196.97	80 Hg Mercury 200.59	81 Tl Thallium 204.38	82 Pb Lead 207.2	83 Bi Bismuth 208.98	84 Po Polonium 209	85 At Astatine 210	86 Rn Radon 222
87 Fr Francium 223	88 Ra Radium 226.025	89 Ac # Actinium 227	104 Rf Rutherfordium 261	105 Db Dubnium 262	106 Sg Seaborgium 263	107 Bh Bohrium 262	108 Hs Hassium 265	109 Mt Meitnerium 266	110 Unn 269	111 Uuu 272	112 Uub 277						

Metals ← → Nonmetals

Lanthanides: *	58 Ce Cerium 140.12	59 Pr Praseodymium 140.908	60 Nd Neodymium 144.24	61 Pm Promethium 145	62 Sm Samarium 150.36	63 Eu Europium 151.96	64 Gd Gadolinium 157.25	65 Tb Terbium 158.925	66 Dy Dysprosium 162.50	67 Ho Holmium 164.93	68 Er Erbium 167.26	69 Tm Thulium 168.93	70 Yb Ytterbium 173.04	71 Lu Lutetium 174.97
Actinides: #	90 Th Thorium 232.038	91 Pa Protactinium 231.036	92 U Uranium 238.029	93 Np Neptunium 237	94 Pu Plutonium 244	95 Am Americium 243	96 Cm Curium 247	97 Bk Berkelium 247	98 Cf Californium 251	99 Es Einsteinium 252	100 Fm Fermium 257	101 Md Mendelevium 258	102 No Nobelium 259	103 Lr Lawrencium 260

Figure 1.2 The Periodic Table of Elements

Practice

Identify the following elements as a metal, nonmetal, or metalloid:

9. Boron
10. Carbon
11. Gold
12. Lead
13. Selenium

LESSON

Chemical Composition: Atoms, Molecules, and Ions

The atom is central to chemistry. Atoms combine to make up the elements, molecules, and compounds that form the basis of chemistry. An understanding of the basic properties of atoms, and how they combine together to form molecules and compounds, will be incredibly helpful as we continue to develop the tools that unlock chemistry.

Dalton's Atomic Theory

In 1808, John Dalton published *A New System of Chemical Philosophy*, which proposed his hypotheses about the nature of matter. Dalton's atomic theory explained that:

- all elements are made of tiny, indivisible particles called atoms (from the Greek *atomos,* meaning indivisible).
- atoms of one element are identical in size, mass, and chemical properties.
- atoms of different elements have different masses and chemical properties.
- compounds are made up of atoms of different elements in a specific ratio.
- atoms cannot be created or destroyed. They can be combined or rearranged in a *chemical reaction.*

Subsequent experiments, notably those of *J.J. Thomson* (discoverer of the electron), *E. Rutherford* (who established that the atom was made of a dense, central core called a *nucleus*, positively charged by protons, and separated from moving electrons by empty space), and others such as *A. Becquerel* and *Marie Curie* (on the spontaneous disintegration of some nucleus with the emission of particles and radiation), were necessary, however, to complete the understanding of atoms.

The *atomic weight* (or mass) of an element is given by the weighted average of the isotopes' masses. Isotopes are atoms of an element that have different masses.

Scientists were able to formulate three laws based on Dalton's atomic theory:

1. **Law of conservation of mass:** The law of conservation of mass states that mass cannot be created or destroyed. If the mass of the combined reactants is 20 grams, then the mass of the combined products *must* be 20 grams.
2. **Law of definite proportions:** The law of definite proportions is derived from Dalton's fourth hypothesis and states that different samples of the same compound always contain the same proportion by the mass of each element. Water (H_2O) always has a ratio of 2 grams of hydrogen to 16 grams of oxygen regardless of the sample size.
3. **Law of multiple proportions:** The law of multiple proportions is also derived from Dalton's fourth hypothesis and states that if two elements combine to form multiple compounds, the ratio of the mass of one element combined with one gram of the other element can always be reduced to a whole number. Hydrogen can combine with oxygen in two ways: water (H_2O) and hydrogen peroxide (H_2O_2). The ratio of oxygen is 1:2.

Isotopes

Elements are defined by their atomic number. An element's atomic number is the number of protons in the atom and is sometimes written as a subscript of the elemental symbol (i.e., $_{11}Na$). Because the mass number defines the elemental symbol (sodium always has 11 protons and carbon always has 6 protons), the atomic number is frequently omitted.

Also important is the mass number of an element. The *mass number* is the sum of protons and neutrons (in the nucleus) of the atom and is written as a superscript of the element's symbol. For an uncharged atom, the number of protons is equal to the number of electrons.

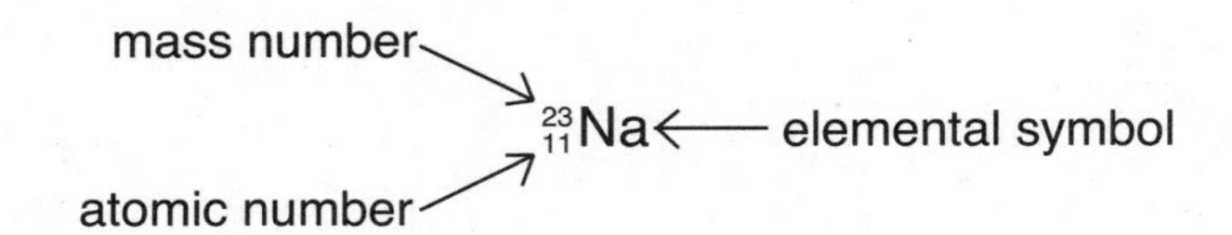

Isotopes are atoms of the same element with the same number of protons (same atomic number) but a different number of neutrons. Isotopes have identical chemical properties (the same reactivity) but different physical properties (i.e., some are radioactive, while others are stable). Consider the three isotopes of hydrogen in Table 2.1.

Table 2.1 Isotopes of Hydrogen

ISOTOPE	SYMBOL	COMMON NAME	NUMBER OF PROTONS	NUMBER OF ELECTRONS	NUMBER OF NEUTRONS	MASS NUMBER
Hydrogen-1	^1H	Hydrogen	1	1	0	1
Hydrogen-2	^2H or D	Deuterium	1	1	1	2
Hydrogen-3	^3H or T	Tritium	1	1	2	3

***Protium or simply proton**

TIPS

Elements usually form +1 cations in group 1A, +2 cations in group 2A, and +3 cations in group 3A. Nonmetals usually form −1 anions in group 7, −2 cations in group 6A, and −3 anions in group 5A.

Practice

Identify the number of protons, neutrons, and electrons in the following isotopes.

1. ^{23}Na
2. ^{99}Tc
3. ^{11}B
4. ^{31}P
5. ^{35}Cl

Ions

As noted earlier, the periodic table is organized in octaves (Groups 1A to 8A). The *octet rule* states that atoms form ions and covalent bonds in order to surround themselves with eight (octet) outer electrons. Notable exceptions are hydrogen (two electrons; duet rule) and group 3 elements (six electrons). They tend to acquire the stability of their closest noble gases in the periodic table either by losing (metals), gaining (nonmetals), or sharing electrons in their valence shell. The valence shell contains the electrons in the outermost energy level.

An *anion* is a negatively charged ion formed when an atom gains one or more electrons. Most anions are nonmetallic. Their names are derived from the elemental name with an *-ide* suffix. For example, when chlorine (Cl) gains an electron, a chloride ion (Cl^-) is formed. Because chlorine is in group 7, it only needs to gain one electron to achieve the octet structure of argon (Ar). An oxygen atom (O) acquires two electrons in its valence shell to form an oxide ion (O^{2-}) that has the same stable electron configuration as neon (Ne).

A *cation* is a positively charged ion and is formed when an atom loses one or more electrons. Most cations are metallic and have the same name as the metallic element. For example, when lithium (Li) loses an electron, a lithium ion (Li^+) is formed. Lithium is in group 1 and needs to lose one electron to acquire the noble gas electron structure of helium (He), the closest noble gas.

Practice

Identify the name and charge of the ions formed from the following elements.

6. Nitrogen
7. Potassium
8. Iodine
9. Magnesium
10. Sulfur

Ionic compounds

Ionic compounds are formed by the combination of a positively charged cation (usually a metal) and a negatively charged anion (usually a nonmetal). The attractive electrostatic force between a cation and an anion is called an ionic bond. Common table salt, sodium chloride (NaCl), is an example of an ionic compound.

Molecules

A molecular compound is formed when two nonmetals combine to form a covalent bond. Covalent bonds are the type of bonds formed when two atoms *share* one or more pairs of electrons to achieve an octet of electrons. A polar covalent bond is formed when the atoms unequally share paired electrons.

Electronegativity is the ability of an atom (in a bond) to attract the electron density more than the other atom(s). Electronegativity increases from left to right and from the bottom to the top of the periodic table. Thus, fluorine (F) is the most electronegative element of the periodic table, with the maximum value of 4.0 in the Pauling scale of electronegativity. Metals are electropositive. See Lesson 11 for more on electronegativity.

Formulas and Nomenclature

An essential step in learning chemistry is understanding chemical formulas and how to name compounds. Compounds can be divided into four classifications:

1. Type I: binary ionic compounds
2. Type II: binary ionic compounds where the metal is capable of possessing more than one type of cation
3. Type III: binary covalent compounds
4. Organic compounds

Type I and II Binary Compounds

Type I and II binary compounds are neutral, ionic compounds that contain two parts: a cation and an anion. When a metal is the cation and a nonmetal is an anion, the following rules are used:

- The cation is always listed first and the anion second.
- For cations that possess possible charges (see Figure 2.1), the charge on the ion must be specified by using a Roman numeral in parentheses following the cation.

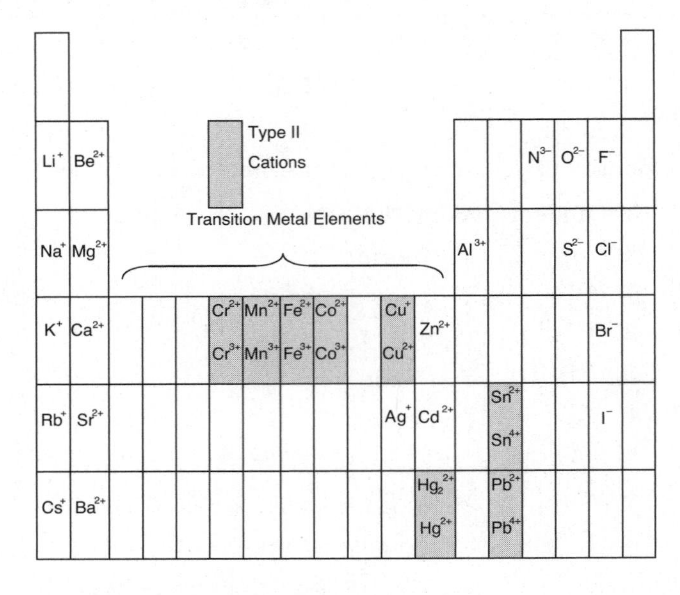

Figure 2.1 Common Cations and Anions

Type I and II compounds are neutral and charges must balance to create a net zero charge.

IONS	COMPOUND	NAME	COMMENT
K^+, Cl^-	KCl	Potassium chloride	Type I
Ca^{2+}, Br^-	$CaBr_2$	Calcium bromide	Type I; two bromides are needed to balance calcium's +2 charge
Mg^{2+}, N^{3-}	Mg_3N_2	Magnesium nitride	Type I
Ag^+, O2$^-$	Ag_2O	Silver oxide	Type I
Fe^{2+}, O^{2+}	FeO	Iron (II) oxide	Type II; need to indicate iron's charge; charges balance
Sn^{4+}, Cl^-	$SnCl_4$	Tin (IV) chloride	Type II

Practice

Name each of the following ionic compounds.

11. PbO
12. $AlCl_3$
13. Fe_2O_3
14. LiF
15. $ZnBr_2$

Write the formula for the following compounds.

16. mercury (II) chloride
17. strontium bromide
18. sodium sulfide
19. manganese (II) oxide
20. cobalt (III) fluoride

Type III: Binary Covalent Compounds

Type III binary compounds are neutral, covalent compounds that contain two or more nonmetals. Type III naming is similar to Type I and II using the following rules:

- The element listed first is named first using the full element name.
- The element listed second is named as if it were an anion.
- A prefix is used to represent the number of atoms because nonmetals can combine in many different ways (see Table 2.3). The prefix *mono-* is not used for the first element.

COMPOUND	NAME
BBr_3	Boron tribromide
CO	Carbon monoxide
N_2O_5	Dinitrogen pentoxide
P_4O_{10}	Tetraphosphorus decoxide

Some covalent molecules use a common name over their systematic name. Examples include H_2O (water), NH_3 (ammonia), CH_4 (methane), and BH_3 (borane).

Practice

Name each of the following covalent compounds.

21. N_2O_4
22. P_2O_5
23. NO

Write the formula for the following compounds.

24. nitrogen trioxide
25. iodine trichloride
26. carbon tetrachloride

Table 2.3 Numerical Prefixes Used in Chemical Names

PREFIX*	NUMBER	PREFIX*	NUMBER
mon(o)-	1	hex(a)-	6
di-	2	hept(a)-	7
tri-	3	oct(a)-	8
tetr(a)-	4	non(a)-	9
pent(a)-	5	dec(a)-	10

***To avoid awkward pronunciations, the "o" or "a" of a prefix is usually dropped if the element's name begins with a vowel.**

Polyatomic Ions

Polyatomic ions are ions that contain more than one atom (see Table 2.2). These ions can replace one or both ions in Type I or II ionic compounds and have special names. However, the oxyanions (the ions containing oxygen) have a systematic naming structure. When two oxyanions of an element are present, the anion with the larger number of oxygen atoms is given the suffix *-ate* (i.e., sulfate, SO_4^{2-}, and nitrate, NO_3^-), and the anion with the smaller number of oxygen atoms is given the suffix *-ite* (i.e., sulfite, SO_3^{2-}, and nitrite, NO_2^-). When more than two oxyanions exist in a series, the prefixes *hypo-* (less than) and *per-* (more than) are used, as in chlorine oxyanions.

IONS	COMPOUND	NAME	COMMENT
NH_4^+, Cl^-	NH_4Cl	Ammonium chloride	Type I
Na^+, SO_4^{2-}	Na_2SO_4	Sodium sulfate	Type I
Ca^{2+}, OH^-	$Ca(OH)_2$	Calcium	Type I; hydroxide polyatomic ions are a unit and need parentheses if more than one is needed
Pb^{+2}, NO^{3-}	$Pb(NO_3)_2$	Lead (II) nitrate	Type II

Table 2.2 Common Polyatomic Ions

FORMULA	NAME	FORMULA	NAME
NH^{4+}	Ammonium	PO_4^{3-}	Phosphate
Hg_2^{2+}	Mercury (I)	ClO^-	Hypochlorite
		ClO_2^-	Chlorite
OH^-	Hydroxide	ClO_3^-	Chlorate
CN^-	Cyanide	ClO_4^-	Perchlorate
$CH_3CO_2^-$ or $C_2H_3O2^-$	Acetate	MnO_4^-	Permanganate
$C_2O_42^-$	Oxalate	CrO_4^{2-}	Chromate
CO_3^{2-}	Carbonate	$Cr_2O_7^{2-}$	Dichromate
SO_4^{2-}	Sulfate	HCO_3^{2-}	Bicarbonate (or hydrogen carbonate)
SO_3^{2-}	Sulfite		
NO_3^-	Nitrate	HSO_4^{2-}	Bisulfate (or hydrogen sulfate)
NO_2^-	Nitrite		

Practice

Name each of the following ionic compounds.

27. $CsClO_4$

28. $NaHCO_3$

29. $Fe(NO_3)_3$

Write the formula for the following compounds.

30. mercury (I) chloride

31. copper (II) nitrate

32. calcium oxalate

Acids

Acids are substances that release positive hydrogen ions (H^+) when dissolved in water. Adding one or more hydrogens to an anion requires a different name.

For anions ending in *-ide*, use a *hydro-* prefix and an *-ic acid* ending. For example, what is the chemical name for HCl and HCN?

ANION	ACID	NAME
Cl- (chloride)	HCl	Hydrochloric acid
CN- (cyanide)	HCN	Hydrocyanic acid

For oxyanions (polyatomic ions containing oxygen), use an *-ic acid* ending for polyatomic ions ending in *-ate* and *-ous* ending for polyatomic ions ending in *-ite*. For example, what is the chemical name for H_2SO_4 and H_2SO_3?

ANION	ACID	NAME
SO_4^{-2}(sulfate)	H_2SO_4	Sulfuric acid
SO_3^{-2}(sulfite)	H_2SO_3	Sulfurous acid

Practice

Name each of the following acids.

33. $HClO_4$

34. HBr

35. H_2S

36. H_3PO_4

37. HNO_2

Table 2.4 Nomenclature of Acids

ANION NAME	ACID NAME	EXAMPLE	
_______ide	Hydro_______ic acid	Bromide (Br^-)	Hydrobromic acid (HBr)
hypo_______ite	Hypo_______ous acid	Hypochlorite (ClO^-)	Hypochlorous acid ($HClO_4$)
_______ite	_______ous acid	Chlorite (ClO_2^-)	Chlorous acid ($HClO_2$)
_______ate	_______ic acid	Chlorate (ClO_3^-)	Chloric acid ($HClO_3$)
per_______ate	per_______ic acid	Perchlorate (ClO_4^-)	Perchloric acid ($HClO_4$)

LESSON 3

Measurement and Units

Chemistry is an experimental science. Measuring and calculating the amount of a substance, the temperature of a reaction system, or the pressure of the surroundings are a few ways to help you understand experiments.

Units of Measurement

Quantitative calculations and qualitative interpretations are fundamental to fully grasp the concepts of chemistry. Quantitative values must include a number and a unit. Two common units of measurement are the conventional (English) system and the metric system. The conventional set of units includes inches, feet, miles, gallons, and pounds. These units, although common in the United States, are not used in science or by most of the world. However, the metric system is becoming more common in the United States. The metric system's base-10 units are easier to use and essential for scientific calculations. However, because most readers of this book are more familiar with the conventional system, it will be necessary to convert to and from the metric system.

Table 3.1 Common Prefixes

PREFIX	SYMBOL	FACTOR	EXPONENTIAL
giga	G	1,000,000,000	10^9
mega	M	1,000,000	10^6
kilo	K	1,000	10^3
hecto	H	100	10^2
deca	da	10	101
		1	10^0
deci	D	0.1	10^{-1}
centi	C	0.01	10^{-2}
milli	m	0.001	10^{-3}
micro	μ	0.000001	10^{-6}
nano	n	0.000000001	10^{-9}

The modern metric system uses SI (Système Internationale d'Unités) units that have seven base units (see Table 3.2). All other units are derived from these base units such as area (m^2), volume (m^3 or 1,000 liters), speed (m/s^2), force (kg m/s^2 or newton, N), and energy (kg m^2/s^2 or joule). There are several non-SI units that are frequently used in chemistry.

Table 3.2 Base SI Units

BASE QUANTITY	NAME	SYMBOL	COMMONLY USED NON-SI UNITS
Length	meter	m	cm, nm
Mass	kilogram	kg	g
Time	second	s	hr
Electric current	ampere	A	—
Thermodynamic temperature	kelvin	K	°C, °F
Amount of substance	mole	mol	—
Luminous intensity	candela	cd	—

Uncertainty and Error

Uncertainty expresses the doubt associated with the accuracy of any single measurement. *Accuracy* establishes how close in agreement a measurement is with the accepted value. The *precision* of a measurement is the degree to which successive measurements agree with each other (the average deviation is minimized). *Error* is the difference between a value obtained experimentally and the standard value accepted by the scientific community. Consider the bull's-eye target patterns in Figure 3.1:

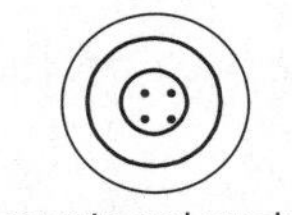

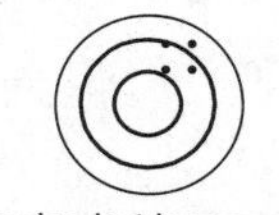

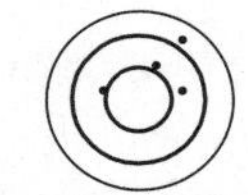

Figure 3.1 Accuracy and Precision

Scientific Notation

Chemists often deal with very large and very small values. As such, it is often inconvenient to describe these numbers using standard decimal notation. Instead, scientific notation is often used. To convert a number to scientific notation, the digit in the largest place value is taken and moved to the ones digit and any significant figures in lower place values are added behind a decimal point. Finally, the number is multiplied by a power of 10 to return the number to its proper value. When written in scientific notation, each digit is significant.

Example

Convert 65,800 and 0.000214 to scientific notation.

STANDARD NOTATION	SCIENTIFIC NOTATION
65,800	6.58×10^4
0.000214	2.14×10^{-4}

Practice

Convert the following numbers to scientific notation.

1. The Earth's mass: 5,974,000,000,000,000,000,000,000 kg
2. The speed of light: 299,000,000 m/s
3. Avogadro's number: 602,200,000,000,000,000,000,000
4. Mass of an electron: 0.000000000000000000000000000000911 kg

Significant Figures

The number of significant figures in any physical quantity or measurement is the number of digits known precisely to be accurate. The last digit to the right is inaccurate. The rules for counting significant figures are as follows:

- All nonzero digits are significant.
- Zeros between nonzero digits are significant figures.
- Zeros that locate the decimal place (placeholder) on the left are nonsignificant.
- Trailing zeros to the right of the decimal point are significant if a decimal point is present.
- All digits used in scientific notation are significant.

NUMBER	NUMBER OF SIGNIFICANT FIGURES	COMMENT
2.34	3	
0.0024	2	Leading zeros are insignificant; number can be rewritten 2.4×10^{-3}.
2005	4	Zeros between nonzero digits are significant.
9400	2	Trailing zeros without a decimal point are nonsignificant.
7.350	4	Trailing zeros with a decimal point are significant.
2.998×10^8	4	Only digits in the number are included, not the exponents.

Practice

Identify the number of significant figures in the following numbers.

5. 5.530

6. 2.4×10^3

7. 0.01215

8. 7.24

9. 121.30

Significant Figures in Calculations

Multiplication and division: The answer will have the same number of significant figures as the least precise number.

$56.2 \times 0.25 = 14.05 = \underline{\underline{14}}$ (0.25 is limiting with two significant figures)

$13.38 \div 12.3 = 1.0878 = \underline{\underline{1.09}}$ (12.3 is limiting with three significant figures)

Addition and subtraction: The answer will have the same number of decimal places as the least precise number.

$12.01 + 1.008 = 13.018 = \underline{\underline{13.02}}$ (12.01 is limiting with two decimal places)

$65.2 - 12.95 = 52.25 = \underline{\underline{52.3}}$ (65.2 is limiting with one decimal place)

When more than one operation is involved in a calculation, note the number of significant figures in each operation, but round *only* the final answer.

$27.43 + 3.32 \times 25.61 = 27.43 + \underline{85.0}252 = 112.4552$
$= \underline{\underline{113}}$
(Three significant figures)

Practice

Solve the following using the correct number of significant figures.

10. $2.4 + 3.69 \times 4.2$

11. $6.022 \times 10^{23} \div 12.0$

12. $(0.24 + 3.25 - 10.2) \div 3.2$

13. $1.5 \times 8.5 - 3.13$

14. $12.13 + 3.3$

Dimensional Analysis (Factor-Label Method)

Dimensional analysis is the method to convert a number from one unit to another using a conversion factor. *Conversion factors* establish a relationship of equivalence in measurements between two different units. Examples of conversion factors are tabulated in Table 3.1 (metric prefixes) and Table 3.3 (common conversion factors).

Within the metric system, the prefix relationship is used (see Table 3.3). It is expressed as a fraction. For instance, for 1 kg = 2.2 lb., the conversion factor is 1 kg/2.2 lb. or 2.2 lb./1 kg.

CONVERT		
FROM	**TO**	**SOLUTION**
185 lbs.	kg	$185 \text{ lbs.} \times \frac{1\text{kg}}{2.205 \text{ lbs.}} = 83.9 \text{ kg}$
4.0 ft.	cm	$4.0 \text{ ft.} \times \frac{12 \text{ in.}}{1 \text{ ft.}} \times \frac{2.54 \text{ cm}}{1 \text{ in.}} = 120 \text{ cm}$
235 mL	qt.	$235 \text{ mL} \times \frac{1 \text{ L}}{10^3 \text{ mL}} \times \frac{1.06 \text{ qt}}{1 \text{ L}} = 0.249 \text{ qt.}$
256 kB	GB	$256 \text{ kB} \times \frac{1 \text{ GB}}{10^6 \text{ kB}} = 2.56 \times 10^{-4} \text{ GB}$ (B is a byte)
2 ft., 3.5 in.	m	$2'3'' = 27.5 \text{ in.} \times \frac{2.54 \text{ cm}}{1 \text{ in.}} \times \frac{1 \text{ m}}{100 \text{ cm}} = .699 \text{ m}$

Practice

Convert the following using the correct number of significant figures.

15. 56 mm to km

16. 3.4 angstroms to centimeters

17. 23.3 oz. to kg

18. 5 ft., 11 in. to m

19. 15.5 qts. to mL

Table 3.3 Common Conversion Factors

LENGTH	MASS	VOLUME
1 in. = 2.54 cm*	1 lb. = 4.536.6 g	1 L = 1.06 qt.
1 m = 1.094 yd.	1 kg = 2.205 lb.	1 gal. = 3.785 L
1 mile = 1.609 km	1 amu = 1.66×10^{-27} kg	1 L = 10^{-3} m^3
1 angstrom = 10^{-10} m		1 mL = 1 cm^3 = 1 c.c.

*Considered to be exact.

Temperature

Temperature is the measure of thermal energy (the total energy of all the atoms and molecules) of a system. The SI unit for temperature is Kelvin, but most scientific thermometers use the centigrade (Celsius) scale. However, most are more familiar with the Fahrenheit scale. Because many chemical calculations require Kelvin temperature, scientists frequently convert from degrees Celsius to Kelvin and from Kelvin to degrees Celsius.

°F to °C: $°C = \frac{5}{9}(°F - 32)$

°C to °F: $°F = \frac{9}{5}(°C + 32)$

K to °C: $°C = K - 273.15$

°C to K: $K = °C + 273.15$

Example:
Room temperature is 70° F. What is 70° F in Celsius and Kelvin scales?

Solution:

$°C = \frac{5}{9}(°F - 32) = \frac{5}{9}(70 - 32) = 21.11 = \underline{21° C}$

$K = °C + 273.15 = 21.11 + 273.15 = 294.26 = \underline{294 K}$

Practice

Convert the following using the correct number of significant figures.

20. 18° C to °F
21. 212° F to K (boiling point of water)
22. 85° F to °C
23. 25° C to K
24. 235 K to °C

LESSON 4

Stoichiometry I: The Mole

Atoms and molecules are too small to count and would require extremely large numbers. Chemists must have an understanding and a unit of measurement to answer "How much is there?" The mole (6.022×10^{23}) is a unit of measurement that describes matter just as a dozen describes a quantity of consumer products.

The Mole

The names and formulas of atoms, molecules, and ions were discussed in Lesson 2, "Chemical Composition: Atoms, Molecules, and Ions." The next step is to define how many atoms, molecules, and/or ions are in a sample. Because these particles are extremely small, a very large number would be required to describe a sample size. Eggs are sold by the dozen (12), soda by the case (24), and paper by the ream (500). Amedeo Avogadro's work on gases helped define a new unit of measurement for chemistry and physics: the mole. A *mole* of a particular substance is defined as the number of atoms in exactly 12 g of the carbon-12 isotope. Experiments established that number to be 6.022142×10^{23} particles.

Avogadro' s number $= N_A = 6.022 \times 10^{23}$ items/mole

In other words, one mole equals 6.022×10^{23} items. The units of the mole are modified depending on the units needed in a calculation. Units could be molecules, atoms, ions, or even everyday items.

Molar Masses

Because the number of grams per element defines the mole, the atomic masses on the periodic table can be given in the units g/mol. Carbon has a molar mass of 12.01 g/mol, and iron has a molar mass of 55.85 g/mol. The *molar mass* of a molecule is calculated by adding all the atomic molar masses.

Molar Calculations

Example:
What is the molar mass of carbon dioxide?

Solution:
Carbon dioxide is CO_2, which contains 1 atom of carbon and 2 atoms of oxygen.

Molar mass of $CO_2 = 12.01 + 2(16.00) = 44.01$ g/mol

Example:
How many moles are there in 62.5 grams of $NaHCO_3$?

Solution:
The molar mass of sodium bicarbonate is needed.

$22.99 + 1.008 + 12.01 + 3(16.00) = 84.008$ g/mol

$$62.5 \text{ g} \left(\frac{\text{mol}}{84.008 \text{ g}}\right) = 0.744 \text{ moles } NaHCO_3$$

Example:
How many atoms are there in 46 grams of carbon?

Solution:
The first step is to use the molar mass to convert grams of carbon to moles of carbon. Once the moles are obtained, Avogadro's number can be used to calculate the number of atoms.

$$46\text{g carbon}\left(\frac{1 \text{ mole}}{12.01 \text{ g carbon}}\right)\left(\frac{6.022 \times 10^{23}\text{atoms}}{1 \text{ mole}}\right)$$

$= 2.3 \times 10^{24}$ atoms

Consider the following:

- One mole of carbon is 12.01 g and contains 6.022×10^{23} atoms.
- One mole of oxygen (O_2) is 32.00 g and contains 6.022×10^{23} molecules.
- One mole of carbon dioxide (CO_2) is 44.01 g and contains 6.022×10^{23} molecules.
- One mole of sodium chloride (NaCl) contains 58.44 g and contains 1 mole of sodium ions (Na^+) and 1 mole of chloride ions (Cl^-).

A molar mass of a molecule can also be called a *molecular mass*.

Practice

Calculate the following:

1. molar mass of nitrogen trioxide
2. molar mass of acetic acid, CH_3CO_2H
3. number of moles of nitrogen trioxide in 67.8 g
4. number of grams in 2.5 moles of acetic acid
5. number of molecules in 36.2 g of acetic acid

Percent Composition

The number of atoms or the mass percent of its elements can describe a compound's composition. The *percent composition* by mass can be calculated by comparing the mass of one mole of each element to the molar mass of the compound.

Note

The *molecular weight* is the sum of the atomic weights of all the atoms in a molecular formula. It is the same number as the molar mass (in grams) without the unit.

Example:
What is the percent composition for each element of glucose, $C_6H_{12}O_6$?

Solution:
First, calculate the mass of each element:

$$\text{carbon} = 6 \text{ mol} \times 12.01 \frac{\text{g}}{\text{mol}} = 72.06 \text{ g}$$

$$\text{hydrogen} = 12 \text{ mol} \times 1.008 \frac{\text{g}}{\text{mol}} = 12.10 \text{ g}$$

$$\text{oxygen} = 6 \text{ mol} \times 16.00 \frac{\text{g}}{\text{mol}} = 96.00 \text{ g}$$

Next, calculate the molar mass (sum of the previous masses):

$$72.06 + 12.10 + 96.00 = 180.16 \text{ g } C_6H_{12}O_6$$

Finally, compare the mass of each element to the molar mass (i.e., a percent is the part over the whole multiplied by 100):

$$\text{mass \% of carbon} = \frac{72.06 \text{ g}}{180.16 \text{ g}} \times 100 = \underline{\underline{40.00\ \%}}$$

$$\text{mass \% of hydrogen} = \frac{12.10 \text{ g}}{180.16 \text{ g}} \times 100 = \underline{\underline{6.716\ \%}}$$

$$\text{mass \% of oxygen} = \frac{96.00 \text{ g}}{180.16 \text{ g}} \times 100 = \underline{\underline{53.29\ \%}}$$

Check: Add the calculated percents to be sure they add to 100 within rounding errors.

$$40.00\% + 6.716\% + 53.29\% \approx 100\%$$

Practice

Calculate the mass percent for each element in each of the following compounds.

6. phosphoric acid, H_3PO_4
7. ethyl alcohol, C_2H_5OH
8. acetone, C_3H_6O
9. iron (III) oxide

Empirical Formula from Composition

An *empirical formula* is the simplest whole-number ratio of atoms in a molecule. Glucose ($C_6H_{12}O_6$) has an empirical formula of CH_2O. For an unknown compound, the empirical formula can be calculated if the mass of each element is known or if the percent of each element is known.

Example:
Determine the empirical formula of a compound that is 78.14% boron and 21.86% hydrogen.

Solution:
Because the mass percent is given, assume 100 g: 78.14 g boron and 21.86 g hydrogen.

Calculate the number of moles of each element:

$$78.14 \text{ g} \times \frac{1 \text{ mol}}{10.811 \text{ g}} = 7.2278 \text{ mol boron}$$

$$21.86 \text{ g} \times \frac{1 \text{ mol}}{1.008 \text{ g}} = 21.687 \text{ mol hydrogen}$$

Because the lowest whole number is needed, divide all moles by the lowest:

$$\text{boron} = \frac{7.2278 \text{ mol}}{7.2278 \text{ mol}} = 1 \text{ mol boron}$$

$$\text{hydrogen} = \frac{21.687 \text{ mol}}{7.2278 \text{ mol}} = 3 \text{ mol hydrogen}$$

Answer: BH_3

Example:
Determine the empirical formula of a compound that has 8.288 g carbon, 1.855 g hydrogen, and 11.04 g oxygen in a sample.

Solution:
Calculate the number of moles of each element:

$$8.288 \text{ g} \times \frac{1 \text{ mol}}{12.011 \text{ g}} = 0.6900 \text{ mol carbon}$$

$$1.855 \text{ g} \times \frac{1 \text{ mol}}{1.008 \text{ g}} = 1.8403 \text{ mol hydrogen}$$

$$11.04 \text{ g} \times \frac{1 \text{ mol}}{16.00 \text{ g}} = 0.69 \text{ mol oxygen}$$

Because the lowest whole-number ratio is required, divide all moles by the lowest. In this situation, all divisors should be 0.69.

$$\text{carbon} = \frac{0.6900 \text{ mol}}{0.69 \text{ mol}} = 1 \text{ mol carbon}$$

$$\text{hydrogen} = \frac{1.8403 \text{ mol}}{0.69 \text{ mol}} = 2.667 \text{ mol hydrogen}$$

$$\text{oxygen} = \frac{0.69 \text{ mol}}{0.69 \text{ mol}} = 1 \text{ mol oxygen}$$

Because the compound must contain whole numbers, $CH_{2.667}O$ is not a valid formula. A multiplier must be used to create a whole-number ratio. In this case, notice that 2.667 is and can be multiplied by 3 to yield 8. The ratio must stay the same, and all elements must be multiplied (or divided in some problems) by the same value.

1 mol carbon × 3 = 3 mol carbon
2.667 mol hydrogen × 3 = 8 mol hydrogen
1 mol oxygen × 3 = 3 mol oxygen
Answer: $C_3H_8O_3$

Practice

Determine the formula for each of the following compositions:

10. 92.26% carbon, 7.74% hydrogen
11. 0.8408 g nitrogen, 0.1208 g hydrogen, 0.360 g carbon, 0.480 g oxygen
12. 5.045 g carbon, 1.693 g hydrogen, 6.72 g oxygen
13. 38.44% carbon, 4.84% hydrogen, 56.72% chlorine

LESSON

5

Stoichiometry II: Chemical Equations

Lesson 4 described "*How much is there*?" and Lesson 5 expands to answer "*How much is supposed to be there*?" Reaction stoichiometry establishes the quantities of reactants (used) and products (obtained) based on a balanced chemical equation.

Chemical Reactions

A *chemical change* or *chemical reaction* can be described by writing a chemical equation. A *chemical equation* uses *chemical symbols* to show what happens during a chemical reaction.

Table 5.1 Basic Types of Chemical Reactions

REACTION TYPE	GENERIC	EXAMPLE
Combination	$A + B \rightarrow C$	$H2 + \frac{1}{2}O_2 \rightarrow H_2O$
Decomposition	$C \rightarrow A + B$	$CaCO_3 \rightarrow CaO + CO_2$
Single displacement (synthesis)	$A + BC \rightarrow B + AC$	$Zn + 2HCl \rightarrow H_2 + ZnCl_2$
Double displacement	$AB + CD \rightarrow AC + BD$	$HCl + NaOH \rightarrow H_2O + NaCl$

Let's look at the electrolysis of water, a decomposition reaction. When water is subjected to an electric current, hydrogen gas and oxygen gas are formed. A chemical equation can be written to show this reaction:

$$H_2O \rightarrow H_2 + O_2$$

The arrow ($\rightarrow$) means *yields*. The *reactants* are a substance that undergoes a change in a chemical reaction (the left side of the arrow). The *products* are a substance that is formed as a result of a chemical reaction (the right side of the arrow). The chemical equation is like a mathematical equation where both sides must be equal. Recall that mass cannot be created or destroyed. Therefore, we must balance the equation. *Balancing equations* is a "trial-and-error" method of equalizing the elements and molecules on both sides of the equation. When balancing an equation, the following guidelines can help:

- Write the unbalanced equation, including the correct formulas, for all reactants and products.
- Compare the number of atoms on the reactants and product(s) sides.
- Balance the elements by changing *the number* of molecules or ions with coefficients. Do *not* change the molecules or ions. The coefficients represent the number of moles of a substance. Always balance the heavier atoms before trying to balance lighter ones such as hydrogen.
- If necessary, continue to rebalance and recheck. Consider reducing the coefficients so that the smallest possible whole numbers are used. Fractions *can* be used in certain circumstances.

For the electrolysis of water, there are 2 hydrogen and 1 oxygen on the reactant side and 2 hydrogen and 2 oxygen on the product side of the equation.

$$H_20 \rightarrow H_2 + O_2$$

	R	P
#H	2	2
#O	1	2

A 2 can be added in front of the water to balance the oxygen:

$$2H_20 \rightarrow H_2 + O_2$$

	R	P
#H	4	2
#O	2	2

However, this change unbalances the hydrogen, so a 2 can be added before the hydrogen gas in the product:

$$2H_20 \rightarrow H_2 + O_2$$

	R	P
#H	4	4
#O	2	2

The equation could also be balanced using fractional values:

$$H_20 \rightarrow H_2 + \tfrac{1}{2}O_2$$

Example:
Write the balanced equation for the combustion of methane and oxygen gas to yield carbon dioxide and water.

1. Write the unbalanced chemical equation:

$$CH_4 + O_2 \rightarrow CO_2 + H_20$$

2. Identify the number of atoms:

$CH_4 + O_2 \rightarrow CO_2 + H_2O$

	R	P
#C	1	1
#O	2	3
#H	4	2

3. Balance the oxygen:

$CH_4 + 2O_2 \rightarrow CO_2 + 2H_2O$

	R	P
#C	1	1
#O	4	4
#H	4	4

4. Recheck (notice that the hydrogen is automatically balanced in this example):

Answer: $CH_4 + 2O_2 \rightarrow CO_2 + 2H_2O$

Practice

Write the balanced equations for the following reactions.

1. $C_2H_6 + O_2 \rightarrow CO_2 + H_2O$
2. $Na + H_2O \rightarrow NaOH + H_2$
3. Ammonium nitrate decomposes to yield dinitrogen monoxide and water.
4. Ammonia reacts with oxygen gas to form nitrogen monoxide and water.
5. Iron (III) oxide reacts with carbon (C) to yield iron metal and carbon monoxide.

Reaction Stoichiometry

Stoichiometry establishes the quantities of reactants (used) and products (obtained) based on a balanced chemical equation. With a balanced equation, you can compare reactants and products, and determine the amount of products that might be formed or the amount or reactants needed to produce a certain amount of a product. However, when comparing different compounds in a reaction, you must *always* compare in moles (i.e., the coefficients). The different types of stoichiometric calculations are summarized in Figure 5.1.

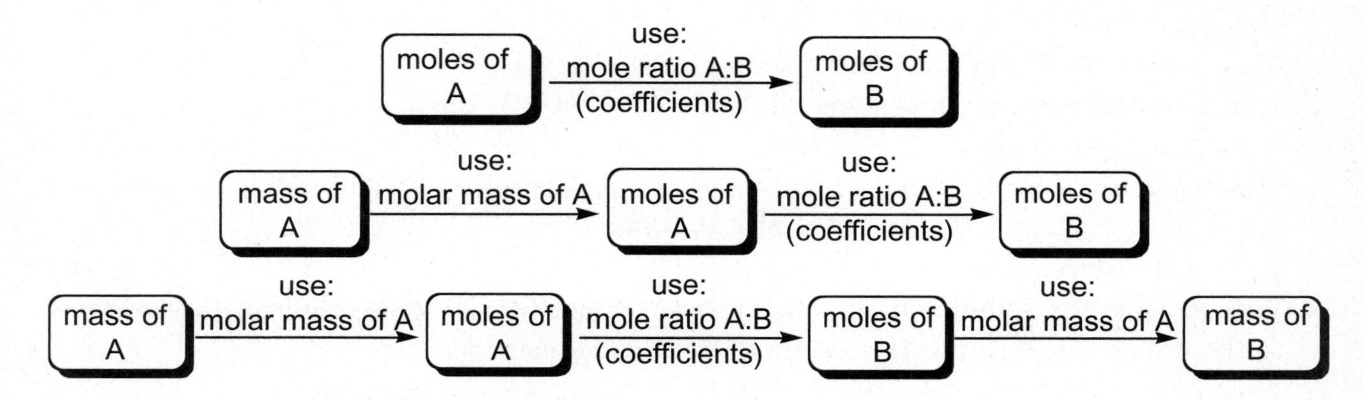

Figure 5.1 Examples of Stoichiometric Calculations

Example:

How many grams of water are produced when 24.3 g of methane is reacted with excess oxygen?

1. Write the unbalanced equation: $CH_4 + O_2 \rightarrow CO_2 + H_2O$

2. Balance the equation: $CH_4 + 2O_2 \rightarrow CO_2 + 2H_2O$

3. Identify the amounts given and needed: $CH_4 + 2O_2 \rightarrow CO_2 + 2H_2O$

   24.3 g → ?

4. Use the molar masses and mole ratios (coefficients) to set up the calculation:

$$24.3\text{ g} \times \frac{1\text{ mol } CH_4}{16.042\text{ g}} \times \frac{2\text{ mol } H_2O}{1\text{ mol } CH_4} \times \frac{18.016\text{ g}}{1\text{ mol } H_2O} = \underline{\underline{54.6\text{ g } H_2O}}$$

molar mass methane   mole ratio   molar mass water

Glucose ($C_6H_{12}O_6$) ferments over time to produce ethanol (C_2H_5OH) and carbon dioxide. How many grams of glucose are needed to produce 100.0 g of ethanol?

5. Write the unbalanced equation: $C_6H_{12}O_4 \rightarrow C_2H_5OH + CO_2$

6. Balance the equation: $C_6H_{12}O_6 \rightarrow 2C_2H_5OH + -CO_2$

7. Identify the amounts given and needed: $C_6H_{12}O_6 \rightarrow 2C_2H_5OH + -CO_2$

   ? → 100.0 g

8. Use the molar masses and mole ratios (coefficients) to set up the calculation:

$$100.0\text{ g} \times \frac{1\text{ mol } C_2H_5OH}{46.068} \times \frac{1\text{ mol } C_6H_{12}O_6}{2\text{ mol } C_2H_5OH} \times \frac{180.156\text{ g}}{1\text{ mol } C_6H_{12}O_6} = \underline{\underline{195.5\text{ g } C_6H_{12}O_6}}$$

molar mass ethanol ratio   mole glucose   molar mass water

Practice

6. Hydrogen gas reacts with carbon monoxide to yield methanol (CH_3OH). How many grams of methanol are formed when 15.6 g of hydrogen react with excess carbon monoxide?
7. How many moles of carbon dioxide are formed in the fermentation of 75 g of glucose?
8. The thermite reaction ($Fe_2O_3 + Al \rightarrow Fe + Al_2O_3$) can be used to ignite solid-fuel rockets or bombs. How much aluminum is needed to react with 10.0 g of Fe_2O_3?

Limiting Reactant

The reagent that is consumed first in a reaction is called the *limiting reactant* or *reagent*. In previous examples and problems, the assumption was that one reactant was in excess and the other was the limiting reactant. However, if a known amount of each reactant is added to a reaction vessel, then the limiting reactant must be calculated. Because the limiting reactant is consumed and limits the amount of products being formed, the easiest method of finding the limiting reactant is to simply calculate the amount of product that *could* be formed from each reactant. The reactant that produces the least amount of product is the *limiting reagent*. The amount of product that the limiting reactant can produce is called the *theoretical yield of the reaction*.

Imagine you are making sandwiches. You have 100 slices of bread and 70 slices of cheese. Each sandwich requires two slices of bread and one slice of cheese. As you make sandwiches, you begin to use up your ingredients. After 50 sandwiches, all the bread is used up, but there are still 20 slices of cheese. In this example, bread is the limiting reagent.

Example:

Identify the limiting reactant and how much ammonia gas can be produced when 8.0 g of nitrogen gas reacts with 8.0 g of hydrogen gas by the use of the Haber process: $3H_2 + N_2 \rightarrow 2NH_3$.

Solution:

1. Identify the amount given and needed:

$3H_2 + N_2 \rightarrow 2NH_3$

$8.0\text{ g} + 8.0\text{ g} \rightarrow ?$

2. Use the molar masses and mole ratios (coefficients) to set up the calculation:

$$8.0\text{ g } H_2 \times \frac{1\text{ mol}}{2.016\text{ g}} \times \frac{2\text{ }NH_3}{3\text{ }H_2} \times \frac{17.034\text{ g}}{1\text{ mol}} =$$

45 g NH_3

$$8.0\text{ g } N_2 \times \frac{1\text{ mol}}{28.02\text{ g}} \times \frac{2\text{ }NH_3}{1\text{ }N_2} \times \frac{17.034\text{ g}}{1\text{ mol}} =$$

9.7 g NH_3

3. The limiting reactant is nitrogen because it can produce the smallest amount of ammonia. The theoretical yield of ammonia is 9.7 g.

Practice

9. Identify the limiting reactant and how much ammonia gas can be produced when 7.2 g of nitrogen gas react with 1.5 g of hydrogen gas by the use of the Haber process: $3H_2 + N_2 \rightarrow 2NH_3$.
10. Identify the limiting reactant and how much carbon dioxide gas can be produced when 15.2 g of methane react with 18.5 g of oxygen gas to produce water and carbon dioxide.
11. Identify the limiting reactant and how much nitric acid can be produced when 60.0 g of nitrogen dioxide react with 18.5 g of water to produce nitric acid and nitrogen monoxide.
12. Identify the limiting reactant and how much aspirin ($C_9H_8O_4$) can be produced when 52.3 g of salicylic acid ($C_8H_6O_3$) react with 25.0 g of acetic acid (CH_3CO_2H): $C_8H_6O_3 + CH_3CO_2H \rightarrow C_9H_8O_4 + H_2O$.

Percent Yield

The *percentage yield* is a ratio of the actual yield of a product over the expected one, known as the *theoretical yield.*

$$\% \text{ yield} = \frac{\text{actual yield}}{\text{theoretical yield}} \times 100$$

Example:
From the previous example, what is the percent yield if only 8.2 g of ammonia is produced?

Solution:

$$\% \text{ yield} = \frac{\text{actual yield}}{\text{theoretical yield}} \times 100 = \frac{8.2 \text{ g}}{9.7 \text{ g}} \times 100$$

$$= 85\% \text{ yield}$$

Practice

Calculate the percent yield for practice problems 9 through 12 if

13. 6.3 g of ammonia were produced from problem 9.
14. 12.4 g of carbon dioxide were produced from problem 10.
15. 51 g of nitric acid were produced from problem 11.
16. 31.0 g of aspirin were produced from problem 12.

LESSON 6

Stoichiometry III: Solutions

Homogeneous solutions are described by how much of a substance is dissolved in a specific solvent. The resulting concentration can be represented by many different units depending on the substance, the solvent, and the solution's potential use.

Concentration of Solutions

The *concentration of a solution* describes the amount of solute that is dissolved in a solvent. The *solute* is a substance that is dissolved in a liquid to form a solution (a homogeneous mixture). The liquid that the solute is dissolved in is the *solvent*. The concentration of a solution can be described in many different ways. The *percent concentration* expresses the concentration as a ratio of the solute's weight (or the volume) over the solution's weight (or the volume). This ratio is then multiplied by 100.

- Weight/volume % = grams of solute/100 mL of solvent
- Volume/volume %= volume of solute/100 volume of *final solution*
- Weight/weight % = grams of solute/100 g of solution

A 20% saline (NaCl or salt) solution contains 20 g of salt per 100 mL of water (20% = 20 g/100 mL × 100). Assuming they are homogeneously mixed, a 45% oil and vinegar solution would be 45 mL per 100 mL. In other words, 100 mL of the oil and vinegar solution would contain 45 mL of oil and 55 mL of vinegar. When dealing with two volumes, the *total volume* must be taken into consideration.

Molarity

A more useful way to describe a concentration is molarity. *Molarity* (M) expresses the number of moles of solute per liter of solution. A 0.1 M NaOH aqueous solution has 0.1 mol of solute (NaOH) in 1 L of water. Because stoichiometric calculations require moles, molarity is more frequently used in calculations.

$$\text{Molarity} = \text{M} = \frac{\text{moles of solute}}{\text{liters of solution}}, \text{ or M} = \frac{\text{mol}}{\text{L}}$$

It is also useful to remember that grams can easily be converted to moles by using the molar mass of a substance.

Example:
What is the molarity of a 20% saline solution?

Solution:

$$\text{M} = \frac{\text{moles of solute}}{\text{liters of solution}}$$

- Calculate the moles:

$$20.\text{ g} \times \frac{1\text{ mole}}{58.44\text{ g}} = .034223$$

- Weight/volume is defined as 100 mL or 0.100 L.
- Calculate M:

$$\text{M} = \frac{0.34223\text{ moles}}{0.100\text{ L}} = 3.4\ \frac{\text{mols}}{\text{L}} = 3.4\text{ M}$$

Practice

1. Calculate the molarity of a solution prepared by adding 45.2 g of solid NaOH to 250 mL of water.
2. Calculate the number of grams of NaOH needed to make a 350 mL solution of a 3.0 M NaOH solution.

Molality

Molality (m) is the number of moles of a solute per kilogram of solvent.

$$\text{Molarity} = \text{M} = \frac{\text{moles of solute}}{\text{kilograms of solution}}, \text{ or}$$

$$\text{M} = \frac{\text{mol}}{\text{kg}}$$

Example:
What is the molality of a mixture made by dissolving 0.50 moles of sugar in 600 g water?

Solution:

$$\text{Molarity} = \text{M} = \frac{\text{mol}}{\text{kg}} = \frac{0.50\text{ moles}}{0.600\text{ kg}} = 0.83\text{ m}$$

(Note: 600 g = 0.600 kg)

Practice

3. Calculate the normality of a solution of 0.56 moles of ethanol in 250 g of water.
4. How many moles of ethanol are needed to make a 1.2 kg, 2.1 m solution?

Normality

Normality (N) is the number of equivalents of the solute per liter of solution. A 1.0 N solution of acid (or base) contains one equivalent of an acid (or base) per liter of solution. A 1.0 M solution of HCl is 1.0 N, but a 1.0 M solution of H_2SO_4 is 2.0 N. Sulfuric acid has two acidic hydrogens, and the molarity is multiplied by a factor of 2. Phosphoric acid (H_3PO_4) is triprotic (having three protons it can donate) and a 1.0 molar solution is 3.0 normal.

Practice

5. Calculate the normality of a solution of 2.56 moles of H_2SO_4 in 250 mL of water.

6. How many moles of H_3PO_4 is needed to make 250 mL of 2.5 N $H_3PO_4(aq)$?

Dilution

The process of adding water to a solution is called *dilution.* Because only the solvent amount is changing, only the total volume and molarity of the solution is changing, not the number of moles of solute. Modifying the molarity equation yields

$$M_1V_1 = M_2V_2$$

The variables M_1 and V_1 represent the molarity and volume of the starting solutions, and M_2 and V_2 represent the final solution. Knowing three of the variables allows the algebraic calculation of the fourth variable.

Example:
What is the final molarity if 75 mL of 12 M concentrated HCl was diluted to 500 mL?

Solution:
So, $M_1 = 12$ M; $V_1 = 75$ mL; $M_2 = ?$; $V_2 = 500$ mL

$$M_1V_1 = M_2V_2 \Rightarrow M_2 = \frac{M_1V_1}{V_2} =$$

$$\frac{(12\text{ M})(75\text{ mL})}{500\text{ mL}} = 1.8\text{ M}$$

How much 3.0 M NaOH is needed to make 350 mL, 0.10 M NaOH solution?

So, $M_1 = 3.0$ M; $V_1 = ?$; $M_2 = 0.10$ M; $V_2 = 350$ mL

$$M_1V_1 = M_2V_2 \Rightarrow V_1 = \frac{M_2V_2}{M_1} =$$

$$\frac{(0.10\text{ M})(350\text{ mL})}{3.0\text{ M}} = 11.7\text{ mL}$$

Practice

7. How much 15.0 M HNO_3 is needed to make 1.5 liters of 1.0 M HNO_3?

8. Calculate the molarity of a solution prepared by combining 450 mL of water with 20 mL of concentrated HCl (12 M).

9. How much 19.2 M NaOH is needed to make 1.5 liters of 1.0 M NaOH?

Part Per Million and Part Per Billion

A common gas and liquid concentration is *parts per million* and *parts per billion.* The "parts" measure the molecules per million or billion:

1 ppm = 1 gram of a compound per every 1,000,000 grams of liquid or gas
1 ppb = 1 gram of a compound per every 1,000,000,000 grams of liquid or gas

These concentration units are typically used when the concentrations are extremely small and it is impractical to use percent or molarity.

Titration

Titration is a chemical technique used to determine the unknown concentration of a known *analyte.* A *titrant,* a known compound of known concentration that will react with the analyte, is added slowly until all the analyte has been reacted. This is known as the *equivalence point* of a titration, where the number of moles of the titrant is equal to the number of moles of analyte. Because the amount of titrant added is known, the number of moles of analyte is known. It is therefore easy to solve for the unknown concentration of analyte.

Usually, a substance called an *indicator,* which changes color when the equivalence point is reached, is added at the beginning of a titration. For example, in an acid-base titration, the indicator phenolphthalein is added, which changes color with changes in pH.

The equation $M_1V_1 = M_2V_2$ can be used to calculate the molarity of the analyte.

Example:
Exactly 45.23 mL of 19.2 M NaOH was titrated to the equivalence point with 75.0 mL of an unkown concentration of HCl. Calculate the molarity of the HCl solution.

Solution:

$$M_1V_1 = M_2V_2 \rightarrow M_1V_1 = M_2V_2 \Rightarrow M_2 = \frac{M_2V_2}{V_2} =$$

$$\frac{(19.2\text{ M})(45.23\text{ mL})}{75.0\text{ mL}} = 11.6\text{ M}$$

Practice

10. Exactly 25.46 mL of 0.125 M NaOH was titrated to the equivalence point with 25.0 mL of an unknown concentration of HCl. Calculate the molarity of the HCl solution.

11. What volume of 2.00 M HNO_3 will react completely with 25.0 mL of 6.00 M NaOH?

12. What volume of 6.00 M KOH will react completely with 25.0 mL of 6.00 M H_2SO_4? (Hint: The sulfuric acid has two protons that can react with the KOH.)

LESSON

7 Solids and Liquids

Unlike gases, solids and liquids are condensed states that contain strong intermolecular forces. These forces result in the properties of solids and liquids.

Intermolecular Forces

Intermolecular forces are the attractive forces that hold molecules and ions together. These forces should not be confused with the *intramolecular forces* that hold the atoms together in a covalent molecule (see Lesson 6, "Molecular Structure"). Intermolecular forces are grouped into five classifications, each supporting the existence of the condensed states of matter: solids and liquids. In addition, these forces can also explain the nonideal behavior of certain gases.

Ion-ion forces are the strongest of the forces and exist between cations and anions in a crystalline structure. A large amount of thermal energy is needed to break up these ions from their orderly, solid structure to a disorderly, liquid structure (e.g. sodium chloride crystals).

Dipole-dipole forces are the forces created by the permanent dipole moment of a polar molecule. For example, acetone ($(CH_3)_2C = O$) has electronegative oxygen that causes a shift in the electron density toward the oxygen. This distribution of electrons leads to the oxygen having a partial negative charge ($\delta -$) and the adjacent carbon having a partial positive charge (δ +). The charges orient themselves like a magnet, with positive to negative ends.

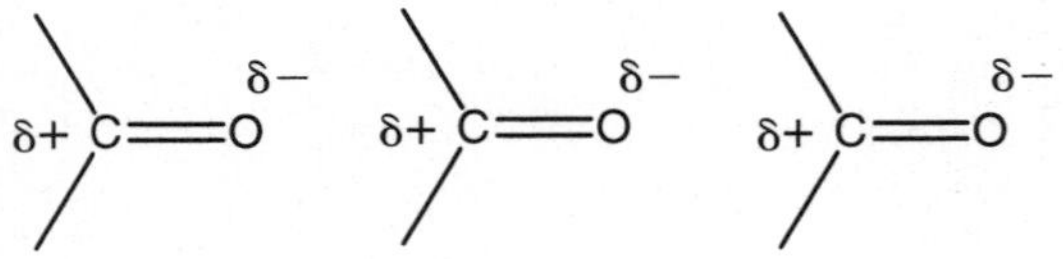

δ+ C=O δ− δ+ C=O δ− δ+ C=O δ−

Ion-dipole forces are a combination of the partial charges of a dipole and the charge of an ion. When table salt (NaCl) dissolves in water, an ion-dipole bond is formed between the sodium and chloride ions and the polar water. Coulomb's Law also explains ion-dipole forces.

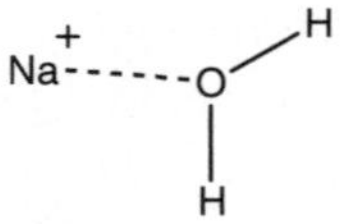

Na^{+}- - - - - -O(H)(H)

Hydrogen bonding is a special intermolecular force that occurs between a hydrogen atom in a very polar bond (N-H, O-H, F-H) and an electronegative nitrogen, oxygen, or fluorine atom in a molecule. Water's molecules are partially held together by hydrogen bonding.

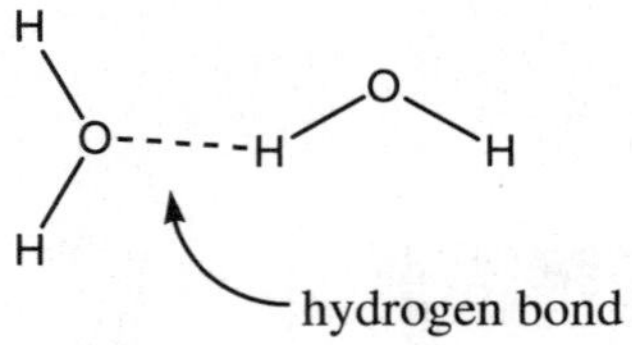

hydrogen bond

Van der Waals forces, also called *dispersion forces*, occur when small, temporary dipoles are formed because of the random motion of electrons. Because electrons are not stationary but constantly in motion, they have the probability of not being equidistant from each other, thus having a balanced atom or molecule. These induced dipoles are weak, attractive forces that occur in all types of matter and exist only momentarily before another induced dipole is formed. The strength of van der Waals forces is related to an atom's polarizability (ease of electron movement). Generally, more polarizable atoms are larger (electrons are farther from the nucleus in a larger orbital and can move -easier).

Example:

Compounds	**Intermolecular forces present**
Methane (CH_4)	van der Waals only (nonpolar molecule)
Lithium bromide	Ion-ion (ionic compound) van der Waals
Methyl alcohol (CH_3OH)	Dipole-dipole (polar molecule) Hydrogen bonding (hydrogen bonded to oxygen) van der Waals
Carbon dioxide (CO_2)	van der Waals only (linear, nonpolar molecule)

Intermolecular forces can also exist between different compounds:

Compounds	**Intermolecular forces present**
HBr and NO_3^-	Dipole-dipole Ion-dipole van der Waals
NO_2 and NH_3	Dipole-dipole Hydrogen bonding van der Waals

Practice

What types of intermolecular forces exist among molecules in each of the following?

1. SO_2
2. $MgCl_2$
3. HF
4. benzene (C_6H_6)

What types of intermolecular forces exist among molecules in each of the following pairs?

5. HF and H_2O
6. PH_3 and H_2S
7. CO_2 and NaOH

Intermolecular forces can also be used to predict the relative boiling and melting point of a compound. When comparing similar compounds, the one with the greater intermolecular forces has a higher boiling and melting point.

Example:
Which compound of the following pairs would you expect to have the higher melting point?

Compounds	Higher melting point (also higher boiling point)
Ne and Kr	Kr; only van der Waals forces are present in the pair and Kr is the largest atom.
NH_3 and NCl_3	NH_3; both are polar molecules, but only ammonia (NH_3) has hydrogen bonding.
CO_2 and NO_2	NO_2; carbon dioxide is linear and nonpolar (only van der Waals forces), and nitrogen dioxide is bent and polar.

Practice

Which compound of the following pairs would you expect to have the highest melting point?

8. CH_3CH_2OH or CH_3OCH_3
9. SO_2 or SO_3
10. HF or HBr
11. Xe or Ar

Heterogeneous mixture: A system of two or more substances (elements or compounds) that have distinct chemical and physical properties. Examples include mixtures of salt and sand, oil and water, crackerjacks, and dirt.

Homogeneous mixture (or solution): A system of two or more substances (elements or compounds) that are interspersed, such as the gases making up the air or salt dissolved in water. The individual substances have distinct chemical properties and can be separated by physical means.

Practice

Identify the following as an element, compound, heterogeneous mixture, or homogeneous mixture:

12. Dry oatmeal
13. Tap water
14. Plutonium
15. Italian oil and vinegar salad dressing
16. Crystal of kosher table salt
17. Gasoline
18. Carbonated soda

Physical Versus Chemical Change

A *physical change* of a substance does not change its chemical composition. The boiling or freezing of water is an example of a physical change. It does not matter if H_2O is a solid (ice), liquid, or gas (steam); it is still water. Popcorn popping is another example of a physical change. The oil heats the water in the popcorn kernel and converts the water into steam. The liquid water changing to steam causes the "explosion" and opening of the kernel, but no chemical change has occurred.

A chemical change or reaction is a process where one or more substances are converted into one or more new substances. The burning logs in a campfire are a chemical change that converts the wood (carbohydrates, etc.) into carbon dioxide, water, and other substances. The roasting of marshmallows over the campfire quickly becomes a chemical change if the marshmallows catch fire and the campfire converts its carbohydrates into carbon.

Practice

Identify the following as a physical change or chemical change:

19. Mixing cake batter
20. Baking a cake
21. Dissolving sugar in tea
22. Burning cookies in the oven
23. Cooking a turkey

Colloids

Colloids are stable mixtures in which particles of rather large sizes (ranging from 1 nm to μm) are dispersed throughout another substance. Aerosol (liquid droplets or solid particles dispersed in a gas) such as fog can scatter a beam of light (Tyndall effect).

Colligative Properties

Colligative properties are solution properties that vary in proportion to the solute concentration and depend *only* on the number of solute particles. This section covers a few solubility laws based on colligative properties.

Henry's Law (a gas-liquid solution) states that the solubility of an ideal gas (C) in mol per liter is directly proportional to the partial pressure *(p)* of the gas relative to a known constant (*k*) for the solvent and gas at a given temperature.

$$C = kp$$

Raoult's Law (a solid-liquid solution) says that the vapor pressure of an ideal solution (p_{total}) is directly proportional to the partial vapor pressure (p_A) of the pure solvent times the mole fraction (X_A = moles of solute per moles of solute and solvent) of the solute.

$$p_{total} = X_A p_A$$

Raoult's Law (a liquid-liquid solution) states that the vapor pressure of an ideal solution of two liquids (p_{total}) is directly proportional to the vapor pressures (p_A° and p_B°) of the pure liquids, the mole fractions of the liquids (X_A and X_B), and the partial vapor pressure (p_A and p_B) of the liquids above the solution.

$$p_{total} = X_A p_A^\circ + X_B p_B^\circ + p_A + p_B$$

Example:
Use Henry's Law to calculate the solubility of oxygen (~21% of the atmosphere) in water at STP (273.15 K and 1.00 atm). The Henry's Law constant for $O_2(g)$ at 273.15 K is 2.5×10^{-3} M.

Solution:

- Calculate p: $p = X \times P_{total} = \frac{21\%\ O_2}{100\%}$ (1.00 atm)= 0.21 atm O_2
- Calculate concentration (C): $C = kp = 2.5 \times 10^{-3} \frac{M}{atm}(0.21\ atm) = 5.25 \times 10^{-4}\ M$

Practice

24. Use Henry's Law to calculate the concentration of carbon dioxide (CO_2) in a soda bottle at 25° C if the internal pressure is at 8.0 atm. The Henry's Law constant for $CO_2(g)$ at 25° C is 3.4×10^{-2} M/atm.

25. Use Raoult's Law to calculate the vapor pressure of 1.5 mol of sugar in 6.50 mol of water at 25° C. Water has a vapor pressure of 23.756 mm Hg at 25° C.

Boiling Point Elevation

Solutions containing nonvolatile solutes have higher boiling points than the pure solvent. The *boiling point elevation* (ΔT_b) is directly proportional to the solvent's boiling point elevation constant (K_b) times the molality (m) of the solute in moles per kg of solvent:

$$\Delta T_b = K_b m$$

Freezing Point Depression

Solutions have lower freezing points than the pure solvent. The *freezing point depression* (ΔT_f) is directly proportional to the solvent's freezing point depression constant (K_f) times the molality (m) of the nonelectrolyte solute in moles per kg of solvent:

$$\Delta T_f = K_f m$$

Table 14.1 Common Boiling Point Elevation (K_b) and Freezing Point Depression (K_f) Constants

SUBSTANCE	$K_b\ \frac{°C\ kg}{mol}$	$K_f\ \frac{°C\ kg}{mol}$
H_2O	0.512	1.86
Diethyl ether ($C_2H_5OC_2H_5$)	2.02	1.79
Ethanol (C_2H_5OH)	1.22	2.00
Benzene	2.53	4.90

Example:

Calculate the freezing point of a sugar solution of 0.34 mol of $C_6H_{12}O_6$ in 275 g of water.

Solution:

1. Find m: $m = \frac{1.25\ moles}{1.55\ kg} = 1.2363\ \frac{moles}{kg}$

   (Note: 275 g = 0.275 kg)

2. Find ΔT_f: $\Delta T_f = K_f m = 1.86 \times 1.2363 = 2.3°\ C$
3. The normal freezing point of water is 0° C, so 0° C − 2.3° C = −2.3° C.
4. A practical example of freezing point depression is the use of salt on icy roads. Salt does not "melt" the ice; it just lowers the freezing point.

Example:

Calculate the boiling point of a solution of 1.25 mol of sucrose (table sugar) dissolved in 1,550 g of water.

Solution:

1. Find m: $m = \frac{1.25\ moles}{1.55\ kg} = 0.80645\ \frac{moles}{kg}$

   (Note: 1,550 g = 1.55 kg)

2. Find ΔT_b: $\Delta T_b = K_b m = 0.512 \times 0.80645 = 0.413°\ C$
3. The normal boiling point of water is 100° C, so 100° C + 0.413° C = 100.413° C.

Practice

26. Calculate the boiling point of a solution of 13.4 mol of pentanol (a nonvolatile substance) dissolved in 255 g of diethyl ether. The boiling point of pure diethyl ether is 34.5° C.

27. Calculate the freezing point of a sugar solution of 0.565 mol of $C_6H_{12}O_6$ in 325 g of water.

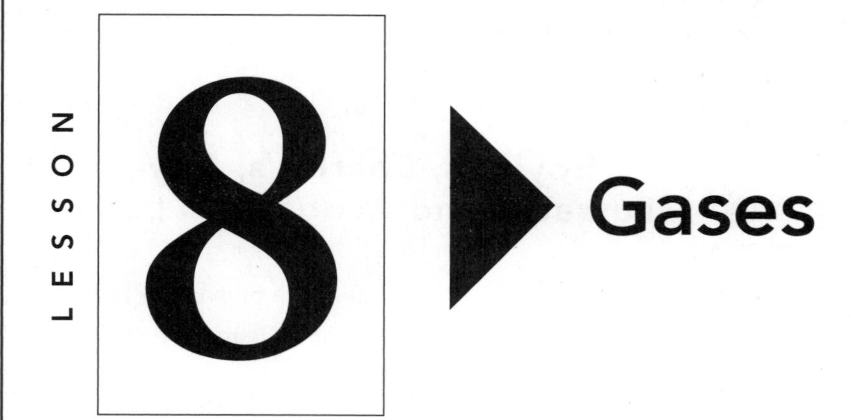

Lesson 8 Gases

Many gases are colorless substances we take for granted. They continually surround us and supply us with oxygen and supply plants with carbon dioxide. Gases' unique property of compressibility allows for quick and observable changes. Changes in pressure, volume, temperature, and other physical properties can be calculated with various ideal gas laws.

Properties of Ideal Gases

We learned in Lesson 2, "Chemical Composition: Atoms, Molecules, and Ions," that gases are fluid, compressible substances. Under most conditions, gases behave according to the following characteristics in the kinetic molecular theory (KMT):

- Gas particles expand to assume the volume and shape of their containers.
- The volume of gas particles is assumed to be negligible.
- Gas particles are in constant motion.
- Gas particles mix evenly and completely when confined in the same container.
- Gas particles collide with each other; they do not attract or repel each other, and they do not exert a force on each other.
- The average kinetic energy of gas molecules is proportional to the temperature (in Kelvin) of the gas. The higher the gas temperature, the higher the kinetic energy.

Note

The most common atmospheric pressure unit used by meteorologists is inches of mercury in Hg (760 mm Hg 5 29.92 in Hg).

Because gases are compressible, they exert pressure on their surroundings. *Pressure* is the force that is exerted over a unit area. For example, the atmosphere exerts a pressure known as *atmospheric pressure*. The Earth's atmosphere is a function of the location and the weather conditions, and it decreases with a higher altitude. The unit of pressure commonly used in chemistry is the atmosphere (atm). The standard atmosphere is 1 atm or a measurement of 760 millimeters of mercury (mm Hg or torr) on a manometer.

$$1 \text{ atm} = 760 \text{ mm Hg} = 760 \text{ torr} = 101{,}325 \text{ Pa (pascal or N/m}^2\text{)}$$

Example:
Covert 830 mm Hg to atmospheres.

Solution:

$$830 \text{ mm Hg} \times \frac{1 \text{ atm}}{760 \text{ mm Hg}} = \underline{\underline{1.09 \text{ atm}}}$$

Practice

Convert the following using the correct number of significant figures.

1. 560 mm Hg to atm
2. 1.23 atm to pascals
3. 2.3 atm to torr

Boyles's, Charles's, Gay-Lussac's, and Avogadro's Laws

Properties of gases that change are pressure (P), temperature (T, in Kelvin), volume (V), and the number of moles (n). Several laws relate these properties:

- **Boyle's Law (at constant temperature):** The volume of a gas (maintained at constant temperature) decreases as its pressure increases ($P \alpha \frac{1}{V}$):

$$P_1V_1 = P_2V_2$$

- **Charles's Law (at constant pressure):** The volume of a gas (maintained at constant pressure) increases directly with an increase in its Kelvin temperature ($V \alpha T$):

$$\frac{V_1}{T_1} = \frac{V_2}{T_2}$$

- **Gay-Lussac's Law (at constant volume):** The pressure of a gas (maintained at constant volume) increases with an increase in its Kelvin temperature ($P \alpha T$):

$$\frac{P_1}{T_1} = \frac{P_2}{T_2}$$

Note

It is important in all the gas law problems that you make sure the answer makes sense. In the previous example, the temperature is increasing, and Charles's Law states that volume increases with an increase in temperature (at constant pressure).

- **Avogadro's Law (at constant T and P):** The volume of gas increases with the number of moles of gas present at constant temperature and pressure. ($V \propto n$)

$$\frac{V_1}{n_1} = \frac{V_2}{n_2}$$

The standard temperature and pressure (STP) condition is 273.15 K and 1 atm (760 torr). One mole (or 6.02×10^{23} particles or molecules) of any gas occupies a volume of 22.4 liters at STP.

Example:
A 2.0 L sample of helium gas is at 25° C. What is the volume of the helium if the sample is heated to 75° C at constant pressure?

Solution:
Temperature and volume changing at constant pressure is Charles's Law. First, all gas law problems require the use of Kelvin temperatures (K = °C + 273.15). The equation is then solved for V_2 (because it is the second temperature that is unknown).

$T_1 = 25 + 273.15 = 298.15$ K $\quad V_1 = 2.0$ L

$T_2 = 75 + 273.15 = 348.15$ K $\quad V_2 = ?$

$$\frac{V_1}{T_1} = \frac{V_2}{T_2} \Rightarrow V_2 = \frac{V_1 T_2}{T_1} = \frac{2.0\text{L} \times 348.15\text{K}}{298.15\text{K}} = 2.3\text{L}$$

Combining Boyle's Law, Charles's Law, and Gay-Lussac's Law forms a combined gas law equation. The combined gas law equation can be used when temperature, pressure, and volume are changing. Essentially, this equation can replace the three individual law equations, and if one of the properties is constant, it can be crossed out and ignored.

Combined Gas Law: $\frac{P_1 V_1}{T_1} = \frac{P_2 V_2}{T_2}$

Example:
On a warm spring day, an automobile tire with a volume of 82.5 L has a pressure of 2.18 atm (32 psi) at 24° C. After driving several hours, the temperature of the tires increases to 49° C and the pressure gauge shows 2.32 atm (34 psi). What is the new volume of the tire?

Solution:
All three variables are changing, so the combined gas law is used to solve the problem.

$P_1 = 2.18$ atm $\quad P_2 = 2.32$ atm

$T_1 = 24° \text{ C} + 273.15 = 297.15$ K $\quad T_2 = 49° \text{ C} + 273.15 = 322.15$ K

$V_1 = 82.5$ L $\quad V_2 = ?$

$$\frac{P_1 V_1}{T_1} = \frac{P_2 V_2}{T_2} \Rightarrow V_2 \frac{P_1 V_1 T_2}{T_1 P_2} =$$

$$\frac{(2.18\text{ atm})(82.5\text{ L})(322.15\text{ K})}{(297.15\text{ K})(2.32\text{ atm})} \; 84.0\text{ L}$$

Practice

4. A child releases a 2.5 L helium balloon that has an internal pressure of 3.4 atm. If a constant temperature is assumed, at what pressure will the balloon be 3.0 L?
5. A 25.0 L beach ball at 760 mm Hg and 25° C is put into a tank of liquid nitrogen at −196° C. What will be the size of the beach ball if the pressure remains constant?
6. At the start of a trip, an automobile tire's pressure reads 32 psi at 25° C. What will be the pressure if friction heats up the tire to 50° C (assume any volume change is negligible)?

Temperature

Temperature is a familiar physical property. We describe foods as hot or cold and check the weather report in the morning to know what the high and low temperatures will be. Temperature is a description of the amount of kinetic energy contained in a gas. The higher the temperature, the faster gas molecules move. Conversely, as a gas cools, its molecules move more slowly. Can a gas get so cold that its molecules stop moving entirely? *Absolute zero* (0 K, −273.15 °C) is the temperature at which particles lose all their kinetic energy. There is no colder temperature.

When examining the proportional relationship between pressure and temperature as well as volume and temperature, scientists found that there was a temperature where the volume of a gas and the pressure of a gas approached zero. These results gave scientists their first clues about absolute zero.

Ideal Gas Law

Under normal conditions, most gases have a similar behavior. The *ideal gas law* equation is used to calculate a variable at a specific point in time. The ideal gas law is a combination of the earlier laws:

Ideal gas law: $PV = nRT$

where $R = \text{Gas constant} = 0.08206 \frac{\text{L} \cdot \text{atm}}{\text{mol} \cdot \text{K}}$

The ideal gas law can also be modified to calculate mass (m) in grams, molar mass (M_m) in moles per gram, or density (d) in grams per liter.

$PVM_m = mRT$ (use when you have mass or need to calculate mass)

$PM_m = dRT$ (use when you have density or need to calculate density)

Example:
It is a mistake to say that humid air is denser than dry air (it's actually the liquid water condensing on you that makes the air "feel" denser). Because water mainly replaces nitrogen gas in the atmosphere, calculate the density of water vapor and nitrogen gas at STP.

Solution:

for N_2: $PM_m = dRT \Rightarrow d = \frac{PM_m}{RT} =$

$$\frac{(1\text{ atm})\left(28.02\frac{\text{g}}{\text{mol}}\right)}{\left(0.08206\frac{\text{L atm}}{\text{mol K}}\right)(273.15\text{K})} = 1.250\frac{\text{g}}{\text{L}}$$

for H_2O: $PM_m = dRT \Rightarrow d = \frac{PM_m}{RT} =$

$$\frac{(1\text{ atm})\left(18.02\frac{\text{g}}{\text{mol}}\right)}{\left(0.08206\frac{\text{L atm}}{\text{mol K}}\right)(273.15\text{K})} = 0.8039\frac{\text{g}}{\text{L}}$$

So nitrogen is 50% denser than water. Note that water is a very small percentage of the atmosphere, and humid air is only a fraction denser than dry air.

Practice

7. How many moles are in a sample of a 50.0 L tank of propane gas at 25° C and 5.60 atm?

8. What is the density of a sample of helium gas at 200° C and 760 mm Hg?

9. An amount of 2.0 g of dry ice (CO_2) is added to an empty balloon. After the dry ice completely sublimes to a gas at 1.0 atm and 25° C, what will be the volume of the balloon?

Dalton's Law of Partial Pressures

In a mixture of gases, individual gases behave independently so that the total pressure is the sum of partial pressures.

$$P_T = p_1 + p_2 + p_3 + \ldots$$

The partial pressures are also directly proportional to the molar amount or percentage of the component.

Example:
Nitrogen is ~78% of the atmosphere and oxygen gas is ~21% of the atmosphere. What is the partial pressure for oxygen gas and nitrogen at STP (760 mm Hg)?

Solution:

$$P_{\text{nitrogen gas}} = 760 \text{ mm Hg} \times 0.78 = 590 \text{ mm Hg}$$

$$P_{\text{oxygen gas}} = 760 \text{ mm Hg} \times 0.21 = 160 \text{ mm Hg}$$

In a balloon, the partial pressure of helium was 3.4 atm and nitrogen was 1.2 atm. What was the total pressure?

Solution:

$$P_{\text{total}} = \text{sum of the partial pressures} = 3.4 \text{ atm} + 1.2 \text{ atm} = 4.6 \text{ atm}$$

Practice

10. An empty 2.5 L diving tank is filled with 1.0 mole of helium and 0.25 moles of oxygen at 25° C. Calculate the total pressure of the tank. (Hint: Use the ideal gas law to calculate the partial pressures.)

11. If the barometer reads 751 mm Hg, what are the partial pressures of oxygen and nitrogen in the atmosphere?

Effusion and Diffusion

Effusion is the passage of a gas through a tiny hole usually into a chamber of lower pressure. Thomas Graham experimentally determined that the rate of effusion is inversely proportional to the square root of the molecular mass. *Graham's law of effusion* states:

$$\frac{\text{rate of effusion for gas A}}{\text{rate of effusion for gas B}} = \frac{\sqrt{\text{molar mass of B}}}{\sqrt{\text{molar mass of A}}}$$

Diffusion is the spread of particles from an area of high concentration to low concentration, such as when a drop of food coloring to spreads throughout a glass of water. Diffusion describes how gases mix together. Graham's Law can be modified to approximate the distance traveled by a gas to mix:

$$\frac{\text{distance traveled by gas A}}{\text{distance traveled by gas B}} = \frac{\sqrt{\text{molar mass of B}}}{\sqrt{\text{molar mass of A}}}$$

Example:
Ammonia gas and hydrogen chloride gas readily come together to form ammonium chloride: $NH_3(g)$ + $HCl(g)$ S $NH_4Cl(s)$. What is the diffusion ratio for these components?

Solution:

$$\frac{\text{distance traveled by gas } NH_3}{\text{distance traveled by gas HCI}} = \frac{\sqrt{\text{molar mass of HCl}}}{\sqrt{\text{molar mass of } NH_3}} = \frac{\sqrt{36.458}}{\sqrt{17.034}} = 1.462$$

That means that the ammonia will travel nearly 1.5 times faster than the HCl gas.

Practice

12. Oxygen and nitrogen are the two major components of the atmosphere. What is the diffusion ratio for oxygen versus nitrogen?

Real Gases

In most instances, the ideal gas law and its variations hold true. However, the behavior of gases does not always follow this simple model. The ideal gas law breaks down in cases of extremely high pressure or when gas particles are attracted or repelled. Real gases must be treated differently with one or more correction factors to be accurate. A common mathematical equation used for real gases is the van der Waals equation. The van der Waals equation corrects the ideal pressure and ideal volume with known constants a and b, respectively, for individual substances.

van der Waals equation:

$$\left[P + a\left(\frac{n}{V}\right)^2\right] \times (V - nb) = nRT$$

The results of an ideal gas calculation are usually a sufficient approximation for most substances and mixtures. The calculations for real gases will not be explored in this text.

LESSON

9 Chemical Equilibria

Stoichiometric calculations assume that reactions progress to completion consuming the limiting reactant. However, because many reaction products can also react to again produce the original reactants or because a reaction does not proceed to completion, chemical equilibrium is established. *Chemical equilibrium* is reached when the amount of products and reactants remain constant.

Equilibrium

Equilibrium is when two opposing reactions occur at the same rate. In chemical equilibrium, the concentrations of the reactants and products remain constant, and no change is observed in the system. Any chemical process will achieve equilibrium over time.

$N_2(g) + 3H_2(g) \rightleftharpoons 2NH_3(g)$ (Haber process)

At equilibrium, there may be greater amounts of products (reaction lies far to the right) or greater amounts of reactants (reaction lies far to the left).

The Equilibrium Constant

For the following general reaction,

$$aA + bB \rightleftharpoons cC + dD$$

where A, B, C, and D are chemical species, and *a*, *b*, *c*, and *d* are their corresponding coefficients, the following equilibrium expression is obtained:

$$K_{eq} = \frac{[C]^c[D]^d}{[A]^a[B]^b}$$

where K is the equilibrium constant.

Example:
Write the equilibrium expression for the Haber process.

1. Write the balanced equation:

$$N_2(g) + 3H_2(g) \rightleftharpoons 2NH_3(g)$$

2. Write the equilibrium expression:

$$K_{eq} = \frac{[NH_3]^2}{[N_2][H_2]^3}$$

Practice

Write the equilibrium expression for the following reactions.

1. $4NH_3(g) + 5O_2(g) \rightleftharpoons 4NO(g) + 6H_2O(g)$
2. $H_2(g) + Br_2(g) \rightleftharpoons 2HBr(g)$
3. $2N_2O_5(g) \rightleftharpoons 4NO_2(g) + O_2(g)$

What about solids and liquids? Pure liquid and solid concentrations do not vary significantly in chemical processes, and their concentrations are always equal to their standard concentrations (usually one). So pure liquids and solids are omitted from equilibrium expressions. Of course, aqueous species are always included in the equilibrium expression.

If a chemical process has a species that is gaseous and one that is aqueous, solid, or liquid, then the gaseous species is usually listed as a partial pressure.

Example:
Write the equilibrium expression for the reaction when gaseous carbon dioxide dissolves in pure water to form carbonic acid.

1. Write the balanced equation:

$$CO_2(g) + H_2O(l) \rightleftharpoons H_2CO_3(aq)$$

2. Write the equilibrium expression:

$$K_{eq} = \frac{[H_2CO_3]}{P_{CO_2}}$$

$CO_2(g)$: a gas, listed as a partial pressure
$H_2O(l)$: a pure liquid, omitted
$H_2CO_3(aq)$: aqueous, listed as its concentration

Practice

Write the equilibrium expression for the following reactions.

4. $PCl_5(s) \rightleftharpoons PCl_3(l) + Cl_2(g)$
5. $H_2SO_4(aq) \rightleftharpoons H_2O(l) + SO_3(g)$
6. $H_2(g) + I_2(g) \rightleftharpoons 2HI(g)$

Solving Equilibrium Problems

The equilibrium constant and equation provide the means to calculate chemical equilibrium. Typically, when the amounts of the reactant or product is given, the equilibrium concentrations can be calculated as long as the equilibrium constant is known.

Equilibrium problems can be solved with the six following steps:

1. Write the balanced equation and equilibrium expression.
2. Identify the initial concentrations of each substance and determine the shift of the equation.
3. Identify the "change" for each substance using a variable (usually x).
4. Add the initial and change to generate the equilibrium amount.
5. Use the equilibrium expression to solve for x.
6. Calculate the equilibrium concentration using the solved x value.

Example:
If the initial concentrations in the Haber process were $[N_2(g)] = 0.90$ M, $[3H_2(g)] = 3.00$ M, and $[2NH_3(g)] = 0$ M, what is the equilibrium concentrations? The equilibrium constant is $6.0 \times 10^{-5}\ M^{-2}$.

Solution:

1. Write the balanced equation and equilibrium expression:

$$N_2(g) + 3H_2(g) \rightleftharpoons 2NH_3(g)$$

$$K_{eq} = \frac{[NH_3]^2}{[N_2][H_2]^3}$$

2. Identify the initial concentrations of each substance and determine the shift of the equation. Identify the "change" for each substance using a variable (usually x). Add the initial and change to generate the equilibrium amount. These three steps can be easily handled as follows:

	$N_2(g)$	$3H_2(g)$	$2NH_3(g)$
Initial	0.90	3.00	0
Change (per mole)	$-x$	$-3x$	$2x$
Equilibrium	$0.90-x$	$3.00-3x$	$2x$

The shift is clearly to the right because no products exist initially. The sign of the change reflects whether we are adding or subtracting to a substance. In this example, because the equilibrium is shifting right, the reactants are disappearing (negative x) and the product is appearing (positive x). The coefficients in the balanced equation corresponding to the number of moles are used as coefficients with the variable x. The initial and change are added to yield the equilibrium concentration.

3. Use the equilibrium expression to solve for x:

$$K_{eq} = \frac{[NH_3]^2}{[N_2][H_2]^3} = 6.0 \times 10^{-5} =$$

$$\frac{[2x]^2}{[0.90 - x][3.00 - 3x]^3}$$

If the K is very small, then the reaction is not expected to proceed far to the right and the "$-x$" and "$-3x$" can be canceled because x will be a small number. This approximation is typical, unless you want to solve the long equation:

$$6.0 \times 10^{-5} = \frac{[2x]^2}{[0.90]\,[3.00]^3}$$

$$[2x]^2 = (6.0 \times 10^{-5})(0.90)(3.00)^3$$

$$x = 0.011$$

4. Calculate the equilibrium concentration using the solved x value.
$[N_2] = 0.90 - x = 0.90 - 0.011 = 0.89$
$[H_2] = 3.00 - 3x = 3.00 - 3(0.011) = 2.97$
$[NH_3] = 2x = 2(0.011) = 0.022$
For equations that do not have small equilibrium constants, the quadratic formula must be used to solve for x:

$$x = \frac{-b \pm \sqrt{b^2 - 4ac}}{2a} \quad \text{where } ax^2 + bx + c = 0$$

Practice

Calculate the equilibrium concentrations for the following processes:

7. For the Haber process where the initial values are $[N_2(g)] = 1.23$ M, $[H_2(g)] = 5.00$ M, and $[NH_3(g)] = 0$ M.
8. For the Haber process where the initial values are $[N_2(g)] = 3.5 \times 10^{-3}$ M, $[H_2(g)] = 1.2 \times 10^{-2}$ M, and $[NH_3(g)] = 0$ M.
9. For $H_2(g) + I_2(g) \rightleftharpoons 2HI(g)$ where the initial values are $[H_2(g)] = 2.5$ M, $[I_2(g)] = 10.0$ M, and $[HI(g)] = 0$ M. $K_{eq} = 1.4 \times 10^{-2}$. Note the quadratic equation must be used to solve for x.

Le Châtelier's Principle

Le Châtelier's principle states that if an equilibrium system is stressed, the equilibrium will shift in the direction that relieves the stress. For the Haber process at equilibrium, if more nitrogen is added, then the process will shift to the right. If nitrogen is removed, then the equilibrium will shift to the left.

Practice

Given the following chemical process for the production of hypochlorous acid,

$$H_2O(g) + Cl_2O(g) \rightleftharpoons 2HClO(g)$$

predict whether the equilibrium will shift to the right or left when

10. water is added.
11. hypochlorous acid is removed.
12. water is removed.
13. $H_2O(g)$ is tripled and $Cl_2O(g)$ is halved.

LESSON

10 Thermodynamics

Thermodynamics is the study of energy and its processes. Energy can exist as *potential energy* (a coiled spring) or *kinetic energy* (an object in motion). *Chemical energy*, a form of potential energy, can be calculated with *calorimetry*, the science of measuring heat, and *enthalpy*, the change of thermal energy of a system. These two components of thermodynamics are essential to understanding heat and thermal energy.

First Law of Thermodynamics

The *first law of thermodynamics* states that energy is neither created nor destroyed. The energy in the universe is constant and can only be converted from one form to another.

Second Law of Thermodynamics

The *second law of thermodynamics* states that there is an increase in entropy (disorder) in a spontaneous process.

Heat Energy and Calorimetry

The *heat of vaporization* is the heat required to evaporate 1 g of a liquid. Water's large heat of vaporization (40.6 KJ/mol) requires large amounts of heat in order to vaporize it into gas. During perspiration, water evaporates from the skin and large amounts of heat are lost. *Heat of fusion* is the heat required to fuse or melt a substance. A *calorie* is defined as the amount of energy required to raise the temperature of 1g of water by 1C. It is equal to 4.18J. One Calorie (uppercase C) is equal to 1000 calories (lowercase c).

Heat capacity is the amount of energy required to raise the temperature of one mole of a substance by 1° C. Water has a high heat capacity, absorbing and releasing large amounts of heat before changing its own temperature. It thus allows the body to maintain a steady temperature even when internal and/or external conditions would increase body temperature. The heat energy of a substance can be calculated using the following equation:

$$q = C \times \Delta T$$

(heat = heat capacity × change in temperature)

However, because the amount of the substance determines how much heat is released or absorbed, it is more useful to use the specific heat capacity or molar heat capacity. The *specific heat capacity* is the heat capacity per gram and is a known value for most substances. The *molar heat capacity* is the heat capacity per mole. The equation can be modified to

$$\text{Energy} = s \times m \times \Delta T$$

(energy or heat = specific heat × mass of substance × change in temperature)

Example:
Calculate the amount of heat released when 150 g of liquid water cools from 100° C (boiling) to 20° C (room temperature).

Solution:
$\text{Energy} = s \times m \times \Delta T = (4.18\ \text{J/g° C})\ (150\ \text{g})\ (20°\ \text{C} - 100°\ \text{C})$

Notice that the change in temperature is final minus initial.

$\text{Energy} = (4.18\ \text{J/g° C})\ (150\ \text{g})\ (-80°\ \text{C}) = -50{,}160\ \text{J} = -50.160\ \text{kJ} = \underline{-50.2\ \text{kJ}}$

The final answer has a negative value that indicates a release of energy.

Table 10.1 Specific and Molar Heat Capacities

SUBSTANCE	SPECIFIC HEAT CAPACITY [J/(G° C)]	MOLAR HEAT CAPACITY [J/(MOL° C)]
$H_2O(l)$	4.18	75.3
$H_2O(s)$	2.03	36.6
C(s)	0.71	8.5
Fe(s)	0.45	25
Cr(s)	0.488	25.4
Al(s)	0.89	24

Practice

1. Calculate the energy released when a 4,570 g piece of hot iron cools from 1,000° C to 20° C.
2. Calculate the energy needed to heat 60 g of aluminum from 100° C to 250° C.
3. Calculate the final temperature of 295 g of water, initially at 30.0° C, if 4,500 joules are added.

Standard Enthalpy of Formation

Chemical reactions can release or absorb energy. *Exothermic* reactions are energy-releasing reactions like most catabolic and oxidative reactions. *Endothermic* reactions are reactions that consume energy in order to take place like anabolic reactions. This energy can be calculated using standard enthalpies of formations. The standard enthalpy of formation (ΔH_f^o) is the energy required to form one mole of a substance from its elements in their standard states. Water can be theoretically formed from its elements, hydrogen and oxygen. Hydrogen is $H_2(g)$ and oxygen is $O_2(g)$ in their standard states. The enthalpy of this exothermic (negative ΔH) reaction is −286 kJ, hence making the standard enthalpy of formation (ΔH_f^o) for liquid water −286 kJ/mol.

$$H_2(g) + O_2(g) \rightarrow H_2O(l) \quad \Delta H = -286 \text{ kJ}$$

The standard enthalpy of reaction (ΔH_{rxn}^o) can be calculated by subtracting the standard enthalpies of formation for each reactant from the standard enthalpies of formation for each product.

$$\Delta H_{rxn}^o = (\Delta H_f^o \text{ products}) - (\Delta H_f^o \text{ reactants})$$

Example:
Calculate the standard enthalpy of reaction for the combustion of methane:

$$CH_4(g) + O_2(g) \rightarrow H_2O(g) + CO_2(g)$$

Solution:
Balance the chemical equation:

$$CH_4(g) + 2O_2(g) \rightarrow 2H_2O(g) + CO_2(g)$$

Table 10.2 Standard Enthalpies of Formations (kJ/mol)

SUBSTANCE	ΔH_f^o	SUBSTANCE	ΔH_f^o	SUBSTANCE	ΔH_f^o
$Al_2O_3(s)$	−1,676	$C_3H_8(g)$	−104	$O_3(g)$	143
$CO(g)$	−110.5	$C_2H_5OH(l)$	−278	$NaOH(s)$	−427
$CO_2(g)$	−393.5	$C_6H_{12}O_6(s)$	−1,275	$NaOH(aq)$	−470
$CH_4(g)$	−75	$H_2O(l)$	−286	$SO_2(g)$	−297
$C_2H_6(g)$	−84.7	$H_2O(g)$	−242	$SO_3(g)$	−396
$C_2H_4(g)$	52	$FeO(s)$	−272	$NO(g)$	90
$C_2H_2(g)$	227	$Fe_2O_3(s)$	−826	$NO_2(g)$	34

NOTE: The $\Delta H_f^o = 0$ for elements in their standard state (i.e., $H_2(g)$, $Fe(s)$, $I_2(s)$, etc.).

Set up the mathematical equation:

$$\Delta H^{o}_{rxn} = (\Delta H^{o}_{f} \text{ products}) - (\Delta H^{o}_{f} \text{ reactants})$$

$$\text{so, } \Delta H^{o}_{rxn} = (2\Delta H^{o}_{f}[H_2O_{(g)}] + \Delta H^{o}_{f}[CO^{2}_{(g)}]) -$$

$$(\Delta H^{o}_{f}[CH_{4(g)}] + 2\Delta H^{o}_{f}[O_{2(g)}])$$

$$\Delta H^{o}_{rxn} = (2(-242) + (-393.5)) - ((-75) + 2(0))$$

[Note: O_2 is oxygen's standard state.)

$$\Delta H^{o}_{rxn} = -877.5 - (-75) = -802.5 = -803 \text{ kJ}$$

As expected, the combustion (burning) of methane is exothermic.

Practice

Calculate the standard enthalpy of reaction for the following reactions (don't forget to balance the equations first!). Also, indicate whether the reaction is exothermic or endothermic.

1. $C_2H_6(g) + O_2(g) \rightarrow H_2O(g) + CO_2(g)$ (combustion of ethane)
2. $H_2O(l) + CO_2(g) \rightarrow C_6H_{12}O_6(s) + O_2(g)$ (photosynthesis)
3. $SO_2(g) + O_2(g) \rightarrow SO_3(g)$
4. $Fe(s) + O_2(g) \rightarrow Fe_2O_3(s)$ (oxidation of iron)

State Functions

Many thermodynamic properties, including enthalpy, are state functions, meaning that the property depends only on the current state of the system, and not the history of the system. This is particularly useful to chemists because the intermediate steps in a chemical reaction, as reactants are converted to products, are not important. Let us consider two examples:

A mountaineer is following the route below from base camp at point A (at 500 m altitude) to the peak at point B (2,700 m altitude). How much altitude has the climber gained when he arrives at point B?

State Functions

Many thermodynamic properties, including enthalpy, are state functions, meaning that the property depends only on the current state of the system, and not the history of the system. This is particularly useful to chemists because the intermediate steps in a chemical reaction, as reactants are converted to products, are not important. Let us consider two examples:

A mountaineer is following the route below from base camp at point A (at 500 m altitude) to the peak at point B (2,700 m altitude). How much altitude has the climber gained when he arrives at point B?

Hess's Law

Many reactions are difficult to calculate with the standard enthalpies of formation or calorimetry. *Hess's Law* states that an enthalpy of a reaction can be calculated from the sum of two or more reactions.

Example:
Calculate the enthalpy for ice, $H_2O(s)$, yielding steam, $H_2O(g)$, given the following equations:

ΔH (kJ)

$H_2O(s) \rightarrow H_2O(l)$ 6.02

$H_2O(I) \rightarrow H_2O(g)$ 44.0

Knowing we need to sum the previous equations to produce $H_2O(s) \rightarrow H_2O(g)$, we get

$H_2O(s) \rightarrow H_2O(I)$ 6.02

$H_2O(I) \rightarrow H_2O(g)$ 44.0

$H_2O(s) \rightarrow H_2O(g)$ 50.0

The yield arrow ($\rightarrow$) acts like an equals sign. Therefore, identical substances on both the right and left sides, regardless of which equations, can be canceled. The sum of the reaction and sum of the enthalpies are added to create the wanted equation and final answer.

Example:
Calculate the enthalpy change for $C(s) + 2H_2(g) \rightarrow CH_4(g)$ given the following equations:

ΔH (kJ)

$C(s) + O_2(g) \rightarrow CO_2(g)$ −393.5

$H_2(g) + \frac{1}{2}O_2(g) \rightarrow H_2O(l)$ −285.8

$CH_2(g) + O_2(g) \rightarrow CO_2(g) + 2H_2O(l)$ −890.3

Solution:
This problem requires more effort. The two rules of Hess's Law must be applied:

- Equations can be reversed if the sign on ΔH is changed.
- Equations can be multiplied by any number to be able to cancel out substances in other equations. The ΔH must also be multiplied by the same value.

We want $C(s) + 2H_2(g) \rightarrow CH_4(g)$. So,

$C(s) + O_2(g) \rightarrow CO_2(g)$ −393.5

$CO_2(g) + 2H_2O(l) \rightarrow CH_4(g) + 2O_2(g)$ + 890.3

(equation is reversed to get CH_4 on the right)

$2[H_2(g) + O_2(g) \rightarrow H_2O(l)]$ 2[−285.8]

(equation is multiplied by 2 to cancel O_2 and H_2O)

$C(s) + 2H_2(g) \rightarrow CH_4(g)$ −74.8 kJ
$(-393.5 + 890.3 + 2(-285.8) = -74.8)$

Practice

1. Calculate the enthalpy change for $P_4(s) + 6Cl_2(g) \rightarrow 4PCl_3(l)$ given

   $P_4(s) + 10Cl_2(g) \rightarrow 4PCl_5(s)\ \Delta H = -1774.0$ kJ

   $PCl_3(l) + Cl_2(g) \rightarrow PCl_5(s)\ \Delta H = -123.8$ kJ

2. Calculate the enthalpy change for $2B(s) + 3H_2(g) \rightarrow B_2H_6(g)$, given

   $4B(s) + 3O_2(g) \rightarrow 2B_2O_3(g)\ \Delta H = -2546$ kJ

   $H_2(g) + \frac{1}{2}O_2(g) \rightarrow H_2O(l)\ \Delta H = -286$ kJ

   $B_2H_6(g) + 3O_2(g) \rightarrow 2B_2O_3(g) + 3H_2(g)\ \Delta H = -2{,}035$ kJ

   $H_2O(g) \rightarrow H_2O_3(g)\ \Delta H = 44$ kJ

Free Energy

Free energy (G) of a system relates the enthalpy (H), temperature in Kelvin (T), and entropy (S) of a system. A negative ΔG means the process is spontaneous, and a positive ΔG means the process is not spontaneous.

$$\Delta G = \Delta H - T\Delta S$$

When is a process spontaneous (i.e., $-\Delta G$)?

If:	**Then:**
$-\Delta H + \Delta S$	spontaneous at all temperatures
$+\Delta H + \Delta S$	spontaneous at high temperatures
$-\Delta H - \Delta S$	spontaneous at low temperatures
$-\Delta H - \Delta S$	not spontaneous at any temperature

LESSON

11 Kinetics/ Reaction Rates

The rate of a reaction describes the speed at which the reactants disappear and products appear. Depending on the chemical reaction and components, reactions rates may be constant or change over time (accelerate or decelerate).

Reaction Rates

The rates of separate chemical reactions can be drastically different. Some reactions occur over a long period of time, such as the conversion of carbon into diamond. Other reactions occur very quickly, such as when sodium carbonate is placed in acetic acid. The *rate of a reaction* is defined as the change in concentration per unit time:

$$\text{Rate} = \frac{\text{change in concentration}}{\text{time}}$$

Rate Laws and Reaction Orders

Rates are affected by three properties:

1. **Temperature:** The rates of reactions increase with the temperature as more collisions among particles occur at higher temperatures.
2. **Particle size:** Smaller particles react faster as they collide, often at any given temperature and concentration.
3. **Concentration:** A high concentration of reacting particles increases the rate of chemical reactions among them.

These factors are incorporated in the rate equation for a chemical reaction. For the reaction $A + B \rightarrow C$, the rate equation is:

$$\text{Rate} = k\,[A]^a[B]^b$$

The k is the rate constant that is dependent on the temperature. The a and b exponents represent the order of the concentrations of A and B. These orders can be proportional to the stoichiometry of the reaction or other experimentally determined value. The concentrations of the reactants can be plotted over time to determine the exact order of each reactant. The overall reaction order is the sum of all the exponents. Several reactions and their rate laws are provided in Table 11.1.

Example:
The following equation represents the oxidation of the bromide ion:

$$5Br^- + BrO_3^- + 6H^+ \rightarrow 3Br_2 + 3H_2O$$

The reaction is first-order bromide ion, first-order bromate ion, and second-order hydrogen ion. What is the rate law and overall order? If the proton concentration quadruples, what is the effect on the rate?

Solution:

$$\text{Rate} = k[Br^-][BrO_3^-][H^+]^2$$

Order = 1 + 1 + 2 = 4th order (summation of the exponents of the rate law equation)

If the proton concentration quadruples, then the relative effect will be 4^2, or 16 times faster.

Practice

The rate for the fixation of ammonia by the Haber process is proportional to the reactant coefficients of the balanced equation:

$$N_2 + 3H_2 \rightarrow 2NH_3$$

Table 11.1 Rate Laws

REACTION	RATE LAW	REACTION ORDER
$CH_3Cl + OH^- \rightarrow CH_3OH + Cl^-$	Rate = $k[CH_3Cl][OH^-]$	2
$2N_2O \rightarrow 2N_2 + O_2$	Rate = $k[N_2O]$	1
$(CH_3)_3CCl + H_2O \rightarrow (CH_3)_3COH + HCl$	Rate = $k[(CH_3)_3CCl]$	1
$CH_4 + Cl_2 \rightarrow CH_3Cl + HCl$	Rate = $k[CH_4][Cl_2]^{1/2}$	1.5

Answer the following questions about the Haber process:

1. What is the rate law of the Haber process?
2. What is the order of the Haber process?
3. How is the rate of consumption of N_2 related to the production of NH_3?
4. How does the rate change if the hydrogen gas concentration is increased by a factor of 2?

Hydrogen peroxide decomposes quickly when exposed to light (notice that hydrogen peroxide is always sold in a dark container.) The rate for the decomposition is first order in H_2O_2.

$$2H_2O_2 \rightarrow 2H_2O + O_2$$

Answer the following questions about the decomposition of hydrogen peroxide:

5. What is the rate law of the decomposition of hydrogen peroxide?
6. What is the order of the decomposition of hydrogen peroxide?

Activation Energy

In the previous lesson, we discussed thermodynamics, which can tell you if a reaction will occur spontaneously (i.e. if the products are lower in energy than the reactants). However, thermodynamics provides no information on how fast this reaction will occur. Extremely exothermic reactions may occur slowly and slightly exothermic reactions may proceed quickly. The combustion of gasoline releases a tremendous amount of energy; however, the fuel is relatively stable and will not spontaneously decompose into carbon dioxide and water without a spark.

The *activation energy* (E_a) of a reaction determines the rate at which a reaction occurs. The activation energy is the minimum amount of energy required to overcome the energy barrier between reactants and products. When the activation energy is high, the reaction is slow. The spark plug in an engine provides the energy to overcome activation barrier of gasoline allowing it to ignite in a controlled way.

Catalysis

Catalysts are species that speed the reaction rate by lowering the activation energy of the reaction. They are not consumed in the reaction.

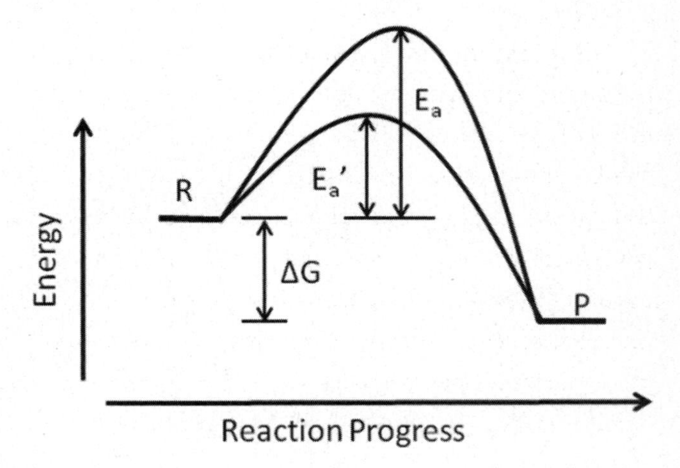

Figure 11.1 A catalyst lowers the activation energy from E_a to E_a', resulting in a faster reaction.

LESSON

12 Aqueous Reactions

Water is the most abundant (and important, besides oxygen) substance on Earth. It is found in large amounts in cells and blood. Water is an excellent solvent and has a high boiling point, high surface tension, high heat vaporization, and low vapor pressure. Three key types of reactions occur in water: precipitation, acid-base, and oxidation-reduction.

Water: A Useful Solvent

Water is a common solvent for many reasons. First, the O-H bonds in H_2O are highly polar, and water forms networks of hydrogen bonds between hydrogen and oxygen atoms of different molecules. The *polarity* of water is designated by the partial positive ($\delta+$) hydrogen and partial negative ($\delta-$) oxygen in Figure 12.1. This polarity allows ions to be soluble and stable in a solution, a requirement for biological activity and when carrying ions through the body. When a substance is dissolved in water, an aqueous solution is formed. An *aqueous solution* is a homogeneous mixture of a substance with water as the solvent.

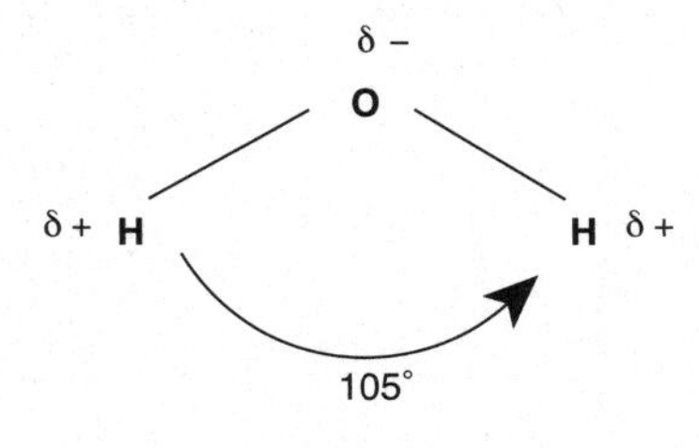

Figure 12.1 Water

One key property of a solution is its electrical conductivity or ability to conduct electricity. When a substance, a *solute*, is dissolved is water, a solvent, ions may or may not be formed. A *strong electrolyte* is formed when the solute completely ionizes (the substance completely separates into ions), such as sodium chloride (a soluble salt), hydrochloric acid (strong acid), or sodium hydroxide (strong base). A *weak electrolyte* is formed when the solute partially ionizes, such as acetic acid (weak acid) or ammonia (weak base). A *nonelectrolyte* is a substance that dissolves in water but does not ionize, such as sugar or alcohol. Most soluble, nonacid organic molecules are nonelectrolytes.

How does a chemist know whether a solute will be a strong or weak electrolyte? Strong acids and bases are strong electrolytes, and weak acids and bases are weak electrolytes. Acids and bases are discussed further in Lesson 15, "Acids and Bases." Soluble salts are also strong electrolytes. Table 12.1 can help determine when a substance is soluble.

What happens if solid lead (II) nitrate, $Pb(NO_3)_2$, is placed in water? According to the solubility table, nitrate compounds are soluble with no exceptions, so lead (II) nitrate would separate into its ions Pb^{2+} and NO_3^-.

What happens if solid silver bromide is placed in water? The solubility table states that bromide compounds are soluble with the exception of silver, lead (II), and mercury (I) ions. Therefore, the silver bromide would be insoluble and remain a solid.

Table 12.1 Solubility Rules for Aqueous Solutions

SOLUBLE IONS	NOTABLE EXCEPTIONS (THESE SALTS ARE INSOLUBLE)
Ammonium (NH_4^+) salts	None
Group I (Li^+, Na^+, K^+, etc.) salts	None
Nitrate (NO_3^-) salts	None
Acetate ($CH_3CO_2^-$) salts	None
Chloride, bromide, iodide (Cl^-, Br^-, I^-) salts	Ag^+: AgCl, AgBr, AgI;
	Pb^{2+}: $PbCl_2$, $PbBr_2$, PbI_2;
	Hg_2^{2+}: Hg_2Cl_2, Hg_2Br_2, Hg_2I_2
Sulfate (SO_4^{2-}) salts	$BaSO_4$, $PbSO_4$, $CaSO_4$, $SrSO_4$, $HgSO_4$,
Bisulfate (HSO_4^-)	None
Most sulfide (S_2^-), carbonate (CO_3^{2-}), phosphate (PO_4^{2-}), chromate (CrO_4^{2-}), and hydroxide (OH^-) salts are insoluble (or only slightly soluble). The exceptions would be these ions combined with group 1 metal ions or ammonium ions. Also, $Ca(OH)_2$, $Sr(OH)_2$, and $Ba(OH)_2$ are marginally soluble.	

Practice

Identify the following substances as a strong electrolyte, weak electrolyte, or nonelectrolyte.

1. sulfuric acid
2. glucose
3. cyanic acid
4. sodium cyanide
5. ammonia

Identify the following as being soluble or insoluble in water. If the substance is soluble, indicate the ions that would be formed upon dissolving in water.

6. chromium nitrate
7. lead (II) iodide
8. barium hydroxide
9. barium sulfate
10. potassium phosphate

Precipitation Reactions

Table 12.1 can also be used to determine the outcome of precipitation reactions. A precipitation reaction occurs when two soluble compounds are mixed and it produces one or more insoluble compounds.

Example:
Write the net ionic equation for the reaction of lead (II) nitrate and sodium chloride.

Solution:
Write the *molecular equation*. The molecular equation shows the reactants and products as molecules. The solubility rules are used to determine if a product is insoluble. Precipitation reactions are examples of a double-displacement reaction. Therefore, the cation of the first molecule (Pb^{2+}) combines with the anion of the second molecule (Cl^-) to produce the first product. The cation of the second molecule (Na^+) combines with the anion of the first molecule (NO_3^-) to produce the second product.

$$Pb(NO_3)_2(aq) + NaCl(aq) \rightarrow PbCl_2(s) + NaNO_3(aq)$$

The molecular equation can be balanced after the products are written:

$$Pb(NO_3)_2(aq) + 2NaCl(aq) \rightarrow PbCl_2(s) + 2NaNO_3(aq)$$

Next, the *ionic equation* is written. The ionic equations show the strong electrolytes (soluble compounds as predicted by the solubility rules) as ions. Of course, solid compounds are not separated.

$$Pb^{2+}(aq) + \underbrace{2NO_3^-(aq)}_{\text{spectator ion}} + \underbrace{2Na^+(aq)}_{\text{spectator ion}} + 2Cl^-(aq) \rightarrow$$

$$PbCl_2(s) + \underbrace{2Na^+(aq)}_{\text{spectator ion}} + \underbrace{2NO_3^-(aq)}_{\text{spectator ion}}$$

Finally, spectator ions (ions which appear on both sides of the chemical equation) are canceled out to reveal the net ionic equations. Spectator ions are ions not involved in the reaction. The *net ionic equation* shows only the species that are directly involved in the reaction (i.e., the spectator ions are not included).

$$Pb^{2+}(aq) + 2Cl^-(aq) \rightarrow PbCl_2(s)$$

Practice

Write the balanced molecular and net ionic equations when the following solutions are mixed together. If no precipitate forms, write "no reaction."

11. $NH_4Cl(aq) + AgNO_3(aq) \rightarrow$
12. $NaOH(aq) + MgCl_2(aq) \rightarrow$
13. lead acetate*(aq)* + sodium sulfate*(aq)* →
14. potassium chloride*(aq)* + lithium carbonate *(aq)* →
15. $K_2S(aq) + Ni(NO_3)_2(aq) \rightarrow$

Acid-Base Reactions

The most general definition of an acid and a base is dependant on water. Svante Arrhenius recognized that certain molecules dissolve in water and produce protons (H^+) and hydroxide ions (OH^-). These substances also react with each other to produce water:

$$H^+ + OH^- \rightarrow H_2O$$

Arrhenius defined *acids* as substances that donate protons in water, and *bases* as substances that donate hydroxide ions in water. However, not all acids and bases contain components that can donate protons or hydroxide ions. Johannes Brønsted and Thomas Lowry defined acids and bases on the proton; acids donate protons and bases accept protons. The Brønsted-Lowry theory of acid-base chemistry is arguably the most widely used for aqueous reactions.

A third definition is based on the valance electron structures developed by Gilbert N. Lewis and does not involve the components of water. The Lewis definition states that substances that can accept electrons in an aqueous solution are acids, and substances that can donate electrons in an aqueous solution are bases. We will revisit Lewis's acid-based theory in Lesson 15, "Acid and Bases."

Practice

Identify the following substances as an acid or a base.

16. HCN
17. HNO_3
18. strontium hydroxide
19. $HClO_4$
20. LiOH

Table 12.2 The Three Definitions of Acid and Bases

	ARRHENIUS	BRØNSTED-LOWRY	LEWIS
Acid	donates H^+ HCl, H_2SO_4, CH_3CO_2H	donates H^+ HCl, H_2SO_4, CH_3CO_2H	accepts electrons H^+, BH_3, $AlCl_3$ (most compounds containing B or Al)
Base	donates OH− NaOH, KOH, $Ca(OH)_2$, $Mg(OH)_2$	accepts H^+ OH^-, NH_3	donates electrons OH^-, NH_3

Notes

The water molecule is a liquid (not aqueous) and does not separate in the ionic equation. HCl is a strong acid and separates into its ions.

Phosphoric acid is a weak acid (i.e., not a strong acid) and not separated in the ionic or net ionic equation.

Writing equations of acid-base reactions follows similar rules as the precipitation reactions. All strong acids are strong electrolytes, and the remaining weak acids are weak electrolytes. The strong acids are listed in Table 12.3 and separate into their ions when placed in solution. However, the weak acids, although soluble, do not separate into their ions. These aqueous weak acids are still labeled *(aq)* but must be treated as a molecule in the ionic and net ionic equations.

Table 12.3 The Six Strong Acids

H_2SO_4 (sulfuric acid)
HNO_3 (nitric acid)
HCl (hydrochloric acid)
HBr (hydrobromic acid)
HI (hydroioic acid)
$HClO_4$ (perchloric acid)

Example:
Write the net ionic equation for the reaction of sodium hydroxide and hydrochloric acid.

- Write the molecular equation:
  $NaOH(aq) + HCl(aq) \rightarrow NaCl(aq) + HOH(l)$ (or H_2O)
- Write the ionic equation:
  $Na^+(aq) + OH^-(aq) + H^+(aq) + Cl^-(aq) \rightarrow Na^+(aq) + Cl^-(aq) + H_2O(l)$
- Write the net ionic equation:
  $OH^-(aq) + H^+(aq) \rightarrow H_2O(l)$

Example:
Write the net ionic equation for the reaction of sodium hydroxide and phosphoric acid.

- Write the molecular equation:
  $3NaOH(aq) + H_3PO_4(aq) \rightarrow Na_3PO_4(aq) + 3HOH(l)$ (or H_2O)
- Write the ionic equation:
  $3Na^+(aq) + 3OH^-(aq) + H_3PO_4(aq) \rightarrow 3Na^+(aq) + PO_4^{3-}(aq) + 3H_2O(l)$
- Write the net ionic equation:
  $3OH^-(aq) + H_3PO_4(aq) \rightarrow PO_4^{3-}(aq) + 3H_2O(l)$

Practice

Write the net ionic equation for the following reactions.

21. cyanic acid*(aq)* + potassium hydroxide*(aq)* $\rightarrow$
22. $Al(OH)_3(s) + HCl(aq) \rightarrow$
23. $CH_3CO_2H(aq) + KOH(aq) \rightarrow$
24. sodium hydroxide*(aq)* + sulfuric acid*(aq)* $\rightarrow$
25. $HNO_2(aq) + Mg(OH)_2(s) \rightarrow$

Oxidation-Reduction Reactions

The *oxidation state* (or oxidation number) for an atom is the number of electrons gained or lost by that atom. Oxidation numbers enable the identification of *oxidized* (an increase in the oxidation number) and *reduced* (a reduction in the oxidation number) elements.

The rules for assigning oxidation state the following:

- The oxidation state of an atom in an element at its standard state is always zero.
  The atoms in $H_2(g)$, $O_2(g)$, $O_3(g)$, $Fe(s)$, and $S_8(s)$ all have an oxidation number of zero.
- The oxidation state of a monatomic (one atom) ion is its charge.
  NaCl: Na is $+1$ and Cl is -1.ZnS: Zn is $+2$ and S is -2.
- Oxygen has a -2 oxidation state in compounds, except hydrogen peroxide (H_2O_2), where O is -1.
  SO_2; O is -2. Consequently, S is $+4$.
  N_2O_5; O is -2. Consequently, N is $+5$.
  H_2O_2; O is -1 and H is $+1$.
- Hydrogen has a $+1$ oxidation state when it is bonded covalently to nonmetals.
  H_2S; H is $+1$. Consequently, S is -2.
  CH_4; H is $+1$. Consequently, C is -4.
- Hydrogen has a -1 oxidation state when it is bonded ionically to metals.
  NaH; H is -1 and Na is $+1$.
  CaH_2; H is -1 and Ca is $+2$.
- Fluorine always has a -1 oxidation state in compounds.
- The sum of the oxidation states must equal zero for compounds and the net charge for polyatomic ions.

Example:
Assign the oxidation state for each element in the following molecules: C_2H_6, $KMnO_4$, and SO_4^{2-}.

C_2H_6: H is $+1$ (covalent compound), so
$6(+1) + 2C = 0$ (compounds are 0), solving for $C = +3$

$KMnO_4$: K is $+1$ (monatomic ion) and oxygen is -2, so
$(+1) + Mn + 4(-2) = 0$ (compounds are 0), solving for $Mn = +7$

SO_4^{2-}: O is -2, so
$S + 4(-2) = -2$ (polyatomic ion = charge), solving for $S = +6$

Practice

Assign the oxidation state for each element in the following molecules.

26. I_2
27. KNO_3
28. H_2CrO_4
29. H_2O_2
30. PF_3

Notes

The redox reaction does not need to be balanced. See Lesson 19 on how to balance redox reactions.

OIL RIG: Oxidation **I**nvolves **L**oss; **R**eduction **I**nvolves **G**ain.

An *oxidation-reduction* reaction, commonly called a *redox reaction*, is characterized by a transfer of electrons. This process is significantly different than the previous two aqueous reactions. In a redox reaction, an electron transfer between the *oxidizing agent* (oxidizes another by accepting its electrons) and the *reducing agent* (reduces another by donating electrons) takes place. Oxidation corresponds to a loss of electrons, and reduction corresponds to a gain of electrons. Sound contradictory? Let us explain using the combustion of methane gas: $CH_4(g) + O_2(g) \rightarrow H_2O(g) + CO_2(g)$. First, the oxidation states must be assigned:

oxidation states:
(for each atom)

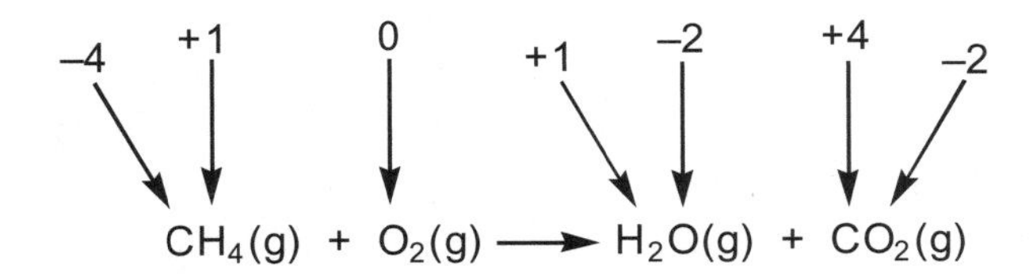

Next, the substances oxidized or reduced can be identified:

- The *species being oxidized* is carbon. Carbon is losing electrons, thus increasing in charge.
- The *species being reduced* is oxygen. Oxygen is gaining electrons, thus decreasing in charge.
- The *oxidizing agent* is O_2. The reactant oxygen gas is the oxidizing agent because it contains the species being reduced.
- The *reducing agent* is CH_4. The reactant methane gas is the reducing agent because it contains the species being oxidized.

The reducing agent *helps* another compound to be reduced, hence being oxidized itself. Likewise, the oxidizing agent *helps* another compound to be oxidized.

Example:
Identify the species being oxidized, the species being reduced, the oxidizing agent, and the reducing agent in the oxidation of methanol (CH_4O) to formaldehyde (CH_2O) with chromic acid:

$$C_2H_6O + H_2CrO_4 \rightarrow C_2H_4O + CrO_2$$

Assign the oxidation states:

oxidation states:
(for each atom)

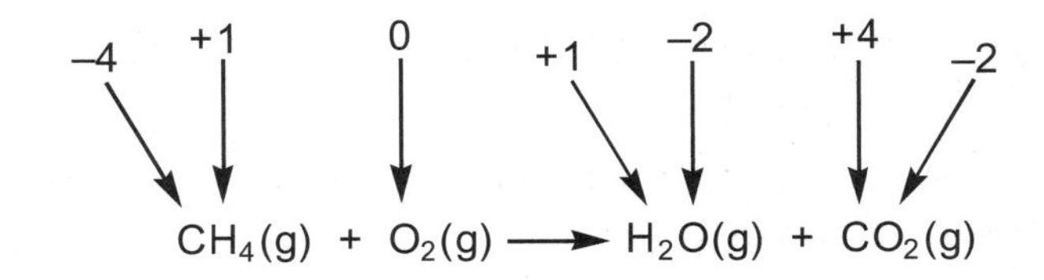

Next, the substances oxidized or reduced can be identified:

- The species being reduced is chromium. Cr is going from a +6 down to a +4 oxidation state.
- The species being oxidized is carbon. C is going from a −2 up to a 0 oxidation state.
- The oxidizing agent is chromic acid (H_2CrO_4), which contains Cr, the species being reduced.
- The reducing agent is methanol (CH_4O). Methanol contains C, the species being oxidized.

Practice

Identify the species being oxidized, the species being reduced, the oxidizing agent, and the reducing agent in the following reactions.

31. $Mg(s) + H_2O(g) \rightarrow Mg(OH)_2(s) + H_2(g)$

32. $8H^+(aq) + 6Cl^-(aq) + Sn(s) + 4NO_3^-(aq) \rightarrow SnCl_6^{2-}(aq) + 4NO_2(g) + 4H_2O(l)$

33. $MnO_4^-(aq) + Fe^{2+}(aq) \rightarrow Fe^{3+}(aq) + Mn^{2+}(aq)$

34. $Na(s) + Cl_2(g) \rightarrow NaCl(s)$

35. $H_2 + O_2 \rightarrow H_2O$

LESSON

13 Acids and Bases

The applications of acid-base chemistry extend beyond what was learned in Lesson 12, "Aqueous Reactions." The *acidity* of an acid solution can be measured by its pH, and *equilibria* can be established between an acid and a base to create a *buffer*.

Acid and Base Definitions

Acids are proton donors (according to Brönsted) or electron acceptors (according to Lewis, this is a more general concept). Strong acids completely dissociate in water, releasing protons (H^+) and anionic conjugate bases. Acids have a sour taste.

Bases are proton acceptors (Brönsted) or electron donors (Lewis). When dissolved in water, strong bases such as NaOH dissociate to release hydroxide ions and sodium cations. Bases have a bitter taste and feel slippery like soap.

Reactions of Acids

When acids react with another substance, the products can be predicted based on solubility rules (see Lesson 12 and Table 12.1) and the four principal acid reactions (see Table 13.1).

Example:
Show the reaction when HCl reacts with $Mg(OH)_2$ (the neutralization of stomach acid with milk of magnesia).

$$2HCl + Mg(OH)_2 \rightleftharpoons MgCl_2 + H_2O$$

Show the reaction when magnesium carbonate reacts with nitric acid.

$$MgCO_3 + 2HNO_3 \rightleftharpoons Mg(NO_3)_2 + [H_2O + CO_2]$$

[from carbonic acid]

Practice

Show the reaction when the following substances react.

1. $H_2SO_4 + NaOH \rightarrow$
2. $Al + HCl \rightarrow$
3. Lithium oxide + nitric acid →
4. Calcium carbonate + HCl →

Autoionization of Water

In pure water, H_2O partially dissociates to H^+ ions (protons) and OH^- ions (hydroxide):

$$H_2O \rightleftharpoons H^+ + OH^-$$

The molar concentration of H^+ equals the molar concentration of OH^-:

$$[H^+] = [OH^-] = 1 \times 10^{-7}\ M$$

In turn, the ion product of water is

$$K_w = [H^+]\,[OH^-] = 1 \times 10^{-14}$$

pH Scale

The pH measures the negative logarithm of the hydrogen ion concentration (in mol/L):

$$pH = -\log [H^+]$$

If pure water has a hydrogen ion concentration of 1×10^{-7} M, the pH of neutral water is 7. A pH of 7 defines a neutral solution. The *pH scale* typically ranges from 0 to 14 with acids in the lower end of the scale

Table 13.1 Reactions of Acids

REACTION	EXAMPLE
Double-Displacement Reactions	
Base + acid → salt + water	$NaOH + HNO_3 \rightarrow NaNO_3 + H_2O$
Metal oxide + acid → salt + water	$CaO + 2HNO_3 \rightarrow Ca(NO_3)_2 + H_2O$
Metal carbonate + acid → salt + carbonic acid	$NaHCO_3 + HCl \rightarrow NaCl + H_2CO_3$ (Note: $H_2CO_3 \rightarrow H_2O + CO_2$)
Single-Displacement Reactions	
Metal + acid → salt + H_2	$Zn + 2HCl \rightarrow ZnCl_2 + H_2$

(smaller than pH 7), whereas bases are at the higher end (greater than pH 7). Some strong acids will result in a negative pH value.

pH of Strong Acids and Bases

Strong acids and bases completely *dissociate* (ionize) in solution. Therefore, the concentration of the solute is the same as the concentration of the $[H^+]$ for acid or $[OH^-]$ for base. A 6.0 M HCl solution produces 6.0 M H^+ ions, and a 3.5 M solution of sodium hydroxide produces 3.5 M of OH^- ions.

Example:
Calculate the pH of a 0.050 M HCl solution.

Solution:
Because HCl is a strong acid, the $[H^+] = 0.05$ M; $pH = -\log[H^+] = -\log(0.050) = \underline{\underline{1.3.}}$

Example:
Calculate the pH of a 0.0030 M NaOH solution.

Solution:
Because NaOH is a strong base, $[OH-] = 0.0030$ M;

$$Kw = [OH^-]\,[H^+] = 1 \times 10^{-14}$$

$$1.0 \times 10^{-14} = [OH^-]\,[H^+] \Rightarrow [H^+] =$$

$$\frac{1.0 \times 10^{-14}}{OH^-} = \frac{1.0 \times 10^{-14}}{0.0030\text{ M}} = 3.3333 \times 10^{-12}\text{M}$$

$$pH = -\log[H^+] = -\log(3.33333 \times 10^{-12}) = 11.477 = 11$$

Practice

5. Calculate the pH of a 0.005 M HNO_3 solution.
6. Calculate the pH of a 0.010 M KOH solution.
7. Calculate the pH of concentrated HCl (12 M).

pH of Weak Acids

Weak acids are weak electrolytes and do not dissociate completely. An equilibrium exists between the reactants and the products, and the equilibrium constant must be taken into account to solve for the pH value. When a weak acid (HA) is dissolved in water, the

Table 13.2 Common Acid Dissociation Constants

ACID	FORMULA*	K_A	PK_A
Hydrofluoric acid	$\underline{H}F$	6.3×10^{-4}	3.2
Formic acid	$HCO_2\underline{H}$	1.82×10^{-4}	3.74
Acetic acid	$CH_3CO_2\underline{H}$	1.78×10^{-5}	4.75
Carbonic acid	$\underline{H}_2CO_3$	4.67×10^{-7} 5.63×10^{-11}	6.35 10.3
Ethanol	$C_2H_5O\underline{H}$	1.0×10^{-16}	16

***The acid hydrogen is underlined.**

conjugate base (A^-) and *conjugate acid* (H^+) are formed. The equilibrium constant for an acid is called the *acid dissociation constant.*

$$HA \rightleftharpoons A^- + H^+$$

[Alternatively written $HA + H_2O \rightleftharpoons A^- + H_3O^+$]

$$K_a = \frac{[A^-][H^+]}{[HA]}$$

and $pK_a = -\log [K_a]$

Example:
What is the pH of a 1.0 M solution of acetic acid?

Solution:

$$CH_3CO_2H \rightleftharpoons CH_3CO^- + H^+$$

$$K_a = \frac{[CH_3CO_2^-][H^+]}{[CH_3COH]} = 1.78 \times 10^{-5}$$

	CH_3CO_2H	CH_3CO^-	H^+
Initial	1.0	0	0
Change	$-x$	$+x$	$+x$
Equilibrium	$1.0 - x$	$+x$	$+x$

$$1.78 \times 10^{-5} = \frac{[x][x]}{[1.0 - x]} = \frac{x^2}{1.0} \Rightarrow 1.78 \times 10^{-5} = x^2$$

$x = 0.0042 = [H^+]$ $\quad$ ($1.0 - x \approx 1.0$ if x is small)

pH $= -\log [0.0042] = \underline{2.4}$

Practice

8. What is the pH of a 0.25 M solution of hydrofluoric acid?

9. What is the pH of a 0.010 M solution of formic acid?

Polyprotic Acids

Substances containing more than one acidic proton are called *polyprotic acids. Diprotic acids* contain two acidic protons, and *triprotic acids* contain three acidic protons. Acid protons dissociate one at a time and have different K_a and pK_a constants. Carbonic acid (H_2CO_3) is a diprotic acid.

$$H_2CO_3 \rightleftharpoons HCO_3^- + H^+$$

$$K_{a1} = 4.67 \times 10^{-7}$$

$$HCO_3^- \rightleftharpoons CO_3^{-2} + H+$$

$$K_{a2} = 5.63 \times 10^{-11}$$

pH of Weak Bases

Problems involving weak bases are treated similarly to the problems with weak acids. Weak bases (B) accept a proton from water (H_2O) to produce a conjugate acid (HB^+) and hydroxide ions:

$$B - H_2O \rightleftharpoons HB^+ + OH^-$$

$$K_b = \frac{[OH^-][HB^+]}{[B][H_2O]} = \frac{[OH^-][HB^+]}{[B]}$$

Since H_2O is the solvent and not a solute it can be removed from the equation.

Practice

10. What is the pH of a 1.25 M solution of ammonia ($K_b = 1.8 \times 10^{-5}$)?

11. What is the pH of a 4.3 M solution of aniline ($K_b = 4.28 \times 10^{-10}$)?

Buffers

A *buffer* is a solution of a weak base and its conjugate acid (also weak) that prevents drastic changes in pH. The weak base reacts with any H^+ ions that could increase acidity, and the weak conjugate acid reacts with OH^- ions that may increase the basicity of the solution.

Phosphate Buffer: $H_2PO_4^-/HPO_4^{-2}$

The principal buffer system inside cells in blood consists of the couple $H_2PO_4^-/HPO_4^{-2}$:

$$H_2PO_4^- \rightleftharpoons HPO_4^{-2} + H^+$$

Neutralization of acid: $HPO_4^{2-} + H^+ \rightarrow H_2PO_4^-$

Neutralization of base: $H_2PO_4^- + OH^- \rightarrow HPO_4^{2-} + H_2O$

The pH of a buffer solution can be calculated by the Henderson-Hasselbalch equation, which states that the pH of a buffer solution has a value close to the value of the weak acid *(pK_a)*:

$$[H^+] = K_a \frac{[acid]}{[base]} \text{ or } pH = pK_a + \log \frac{[base]}{[acid]}$$

Example:

What is the pH of a carbonate buffer solution with 0.65 M H_2CO_3 and 0.25 M HCO_3^-?

Solution:

$$H_2CO_3 \rightleftharpoons HCO_3^- + H^+$$

acid base

$$pH = pK_a + \log \frac{[base]}{[acid]} = 6.35 + \log \frac{[0.25]}{[0.65]} = \underline{5.94}$$

Practice

12. What is the pH of a phosphate buffer solution with 1.35 M $H_2PO_4^-$ and 0.58 M HPO_4^-2 (K_a of $H_2PO_4^-$ is 6.2×10^{-8})?

13. What is the pH of a carbonate buffer solution with 1.36 M H_2CO_3 and 0.61 M HCO_3^-?

Atomic Structure I

Electrons not only orbit around the nucleus of an atom, but they also move to higher and lower orbits also called higher and lower energy levels. These transitions require energy to move to a higher orbit and release energy to move to a lower orbit. The position of an electron can be identified by its unique set of four quantum numbers.

Electromagnetic Spectrum

The *electromagnetic spectrum* represents all types of radiant energy (see Figure 14.1). All radiant energy travels at the speed of light ($c = 2.998 \times 10^8$ m/s). Electromagnetic radiation travels as a wave, and the product of the *wavelength* λ (the distance between two equal points on a wave) and the *frequency* ν (the number of waves that pass a certain point in time) equals the speed of light, or $c = \lambda\nu$, where c = speed of light. Frequency units are in 1/s or Hertz (Hz).

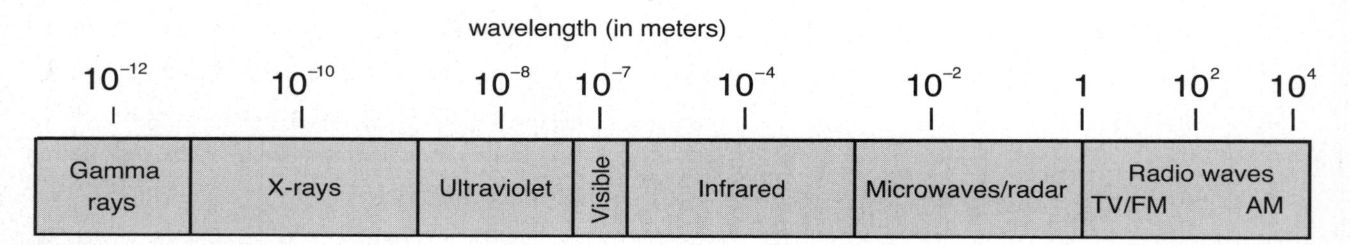

Figure 14.1 Electromagnetic spectrum

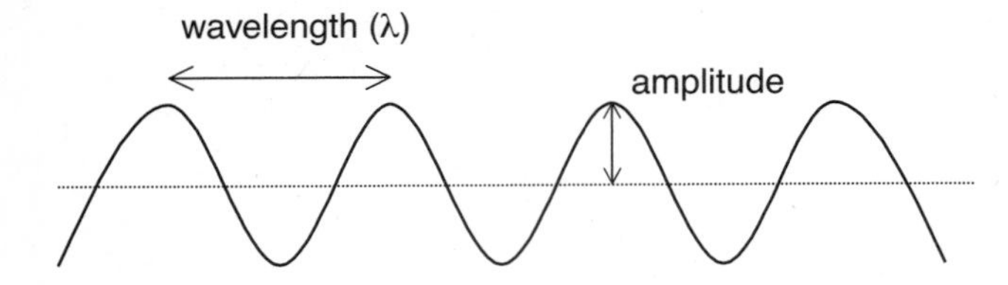

Figure 14.2 Wavelength and amplitude of a wave

Thomas Young described this radiation as a wave, but German physicist Max Planck described this radiation as a "quanta" of energy. Planck discovered that atoms release energy in specific quantities, which he called *quanta*. Albert Einstein was the first to recognize that electromagnetic radiation or light is a combination of quanta theory and wave theory. He suggested that a stream of particles called *photons* travels as a wave through space. Using Planck's theory and constant ($h = 6.626 \times 10^{-34}$ J s), he developed the energy of light or a photon:

$$E_{photon} = h\nu \text{ or simply } E = h\nu$$

Combining these equations gives

$$E = \frac{hc}{\lambda}$$

Example:
Calculate the energy of the photon if the wavelength is 3.62×10^6 nm.

- Convert nm to m (nanometers must be converted to meters because the speed of light is in meters):

$$3.62 \times 10^6 \text{ nm} \times \frac{1 \text{ m}}{10^9 \text{ nm}} = 3.62 \times 10^{-3}\text{m}$$

- Calculate E:

$$E = \frac{hc}{\lambda} = \frac{(6.626 \times 10^{-34}\text{J s})(2.998 \times 10^8 \text{ m/s})}{3.62 \times 10^{-3}\text{m}} =$$

5.49×10^{-23} J

Practice

1. What is the wavelength in nm if the frequency is 1.50×10^{14} Hz?
2. What is the energy of a photon if the frequency is 3.34×10^{12} Hz?
3. What is the frequency if the wavelength is 650 nm?
4. What is the frequency if the energy of the photon is 9.63×10^{-22} J?

Bohr Atom

Niels Bohr's "planetary" model of the hydrogen atom—in which a nucleus is surrounded by orbits of electrons—resembles the solar system. Electrons could be excited by *quanta* of energy and move to an outer orbit (*excited level*). They could also emit radiation when falling to their original orbit (*ground state*). Basic components of the Bohr model include the following:

- **Energy levels:** Energy levels are the volume of space where certain electrons of specific energy are restricted to move around the nucleus. Energy levels consist of one or more orbitals. Energy levels are categorized by the letter n using whole numbers (n = 1, 2, 3, 4 . . .).
- **Orbitals:** An *orbital* describes the probability of finding an electron or pair of electrons in a region of space. These are mathematical functions with specific shapes (s orbitals: spherical; p orbitals: dumbbell, etc.; see Figure 14.2) and restricted zones (called *nodes;* see Figure 14.3). The nodes represent areas where the probability of an electron is zero.

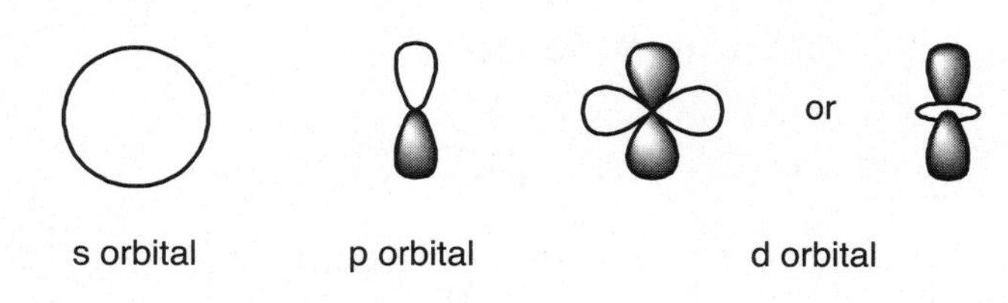

Figure 14.3 Orbital shapes

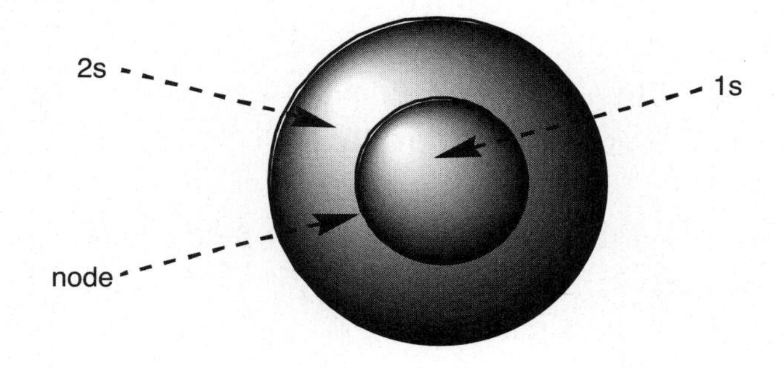

Figure 14.4
1s and 2s orbitals separated by a nodal surface

- **Outer or valence shell:** The *valance shell* is the last energy level containing loosely held electrons. These are the electrons that engage into bonding and are therefore characteristic of the element's chemical properties.

Energy of the Hydrogen Electron

The neutral hydrogen atom has one electron in its ground state or ground level. The ground state is the lowest energy of the atom. The energy levels of an atom are designated by the letter n. Atoms have energy levels beginning with $n = 1$ and go up sequentially in whole numbers ($n = 1, 2, 3,$). When $n = \infty$ the electron separates from the nucleus (see Figure 14.4).

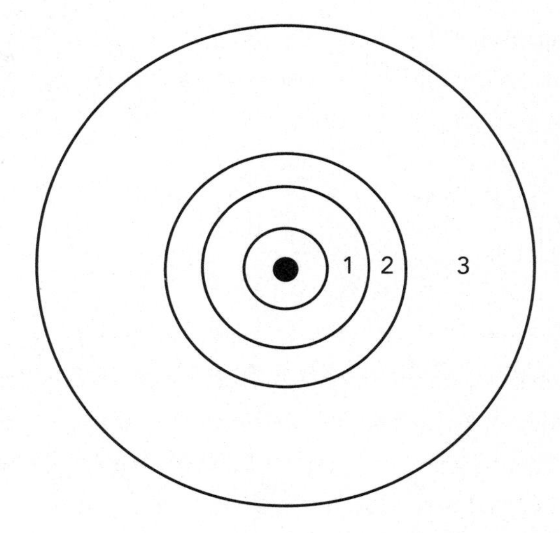

Figure 14.4 The first three energy levels of an atom

When the electron "jumps" to a higher energy level, the electron is in an excited state or excited level. The electron in the excited state returns to the ground state and releases energy equal to the difference in the energy levels. Emission of a photon occurs when excited electrons move to lower energy levels. The chemistry of fireworks is based on the electron movement between the ground state and the excited states. Magnesium has a strong white emission spectrum and lithium has a strong red spectrum. The hydrogen atom energies can be calculated by using the following equation:

$$E = 2.18 \times 10^{-18}\,\text{J}\left(\frac{1}{n_i^2} - \frac{1}{n_f^2}\right)$$

The n_i is the initial energy level and the n_f is the final energy level of the electron. A decrease in the energy level (higher E to lower E) releases energy (minus value) and shows the emission process. However, an increase in the energy level shows the energy required (positive E) to excite the electron.

Example:
What is the energy release when the hydrogen electron goes from n = 5 to n = 2?

$$E = 2.18 \times 10^{-18}\,J\left(\frac{1}{n_i^2} - \frac{1}{n_f^2}\right) = 2.18 \times 10^{-18}\,J$$

$$\left(\frac{1}{5^2} - \frac{1}{2^2}\right) = -4.58 \times 10^{-19}\,J$$

What is the wavelength of the light needed to remove the hydrogen electron from its ground state?

- $n_i = 1$ and $n_f = \infty$ (completely removed from the atom)
- Calculate E:

$$E = 2.18 \times 10^{-18}\,J\left(\frac{1}{n_i^2} - \frac{1}{n_f^2}\right) = 2.18 \times 10^{-18}\,J$$

$$\left(\frac{1}{1_2} - \frac{1}{\infty_2}\right) = 2.18 \times 10^{-18}\,J$$

Note: $\frac{1}{\infty^2} = 0$

- Calculate the wavelength:

$$E = \frac{hc}{\lambda} \Rightarrow \lambda = \frac{hc}{E}$$

$$= \frac{(6.626\ 3\ 10^{234}\ J\ s)(2.998 \times 10^{8}\,m/s)}{2.18 \times 10^{-18}\,J} =$$

$9.11 \times 10^{-8}\,m = 91.1\ nm$

Practice

Calculate the energy and wavelength for the following transitions:

5. n = 1 to n = 5
6. n = 4 to n = 2
7. n = 8 to n = 1
8. n = ∞ to n = 2

LESSON

15 Atomic Structure II

An electron's position in an atom or ion can be described by determining its electron configuration and orbital diagram. These representations of an atom or ion can explain physical and chemical properties of the substance, including magnetic attraction.

Quantum Numbers

Four quantum numbers describe the position and behavior of an electron in an atom: the *principal* quantum number, the *angular* momentum quantum number, the *magnetic* quantum number, and the *electron spin* quantum number. A branch of physics called *quantum mechanics* mathematically derives these numbers through the Schrödinger equation.

Principal quantum number (n): The principle quantum number is the energy level of the electron given the designation n. The value of n can be any integer (1, 2, 3, 4 . . .) and determines the energy of the orbitals.

Angular momentum quantum number (l): The angular momentum is the subshell designation of an electron and describes the shape of the orbital. These subshells are described by the letter l. The possible l values for a particular energy level are 0 to $(n - 1)$. The l values are also given a letter designation. The l value is 0, 1, 2, 3, or 4, and the designation is s, p, d, f, or g.

Magnetic quantum number (m_l): The magnetic quantum number (m_l) describes the orbital's orientation in space. For a given l value, m_l has integer values from $-l$ to $+l$. In other words, for the p subshell ($l = 1$), the m_l values are -1, 0, and $+1$, hence three orbitals.

Electron spin quantum number (m_s): The electron spin quantum number describes the spin of an electron. Magnetic fields have shown that the two electrons in an orbital have equal and opposite spins. The m_s values for these spins are $-\frac{1}{2}$ and $+\frac{1}{2}$.

Example:
Are the following possible sets of quantum numbers? Explain.

(1, 1, 1,); (2, 1, 0, −); (3, 2, 1, 1)

(1, 1, 1,): Not possible. For $n = 1$, l can only be 0, not 1.

(2, 1, 0, $-\frac{1}{2}$): Possible.

(3, 2, 1, 1): Not possible. m_s can only be $-\frac{1}{2}$ or $+\frac{1}{2}$.

Practice

Are the following possible sets of quantum numbers? Explain.

1. (0, 0, 0, $\frac{1}{2}$)
2. (2, 1, 2, $\frac{1}{2}$)
3. (3, 2, −1, $\frac{1}{2}$)
4. (4, 0, 0, $-\frac{1}{2}$)
5. (3, 2, 2, 1)

Table 15.1 Quantum Numbers

N	L	SUBLEVEL DESIGNATION	ML	# OF ORBITALS
1	0	1s	0	1
2	0	2s	0	1
	1	2p	−1, 0, +1	3
3	0	3s	0	1
	1	3p	−1, 0, +1	3
	2	3d	−2, −1, 0, +1, +2	5
4	0	4s	0	1
	1	4p	−1, 0, +1	3
	2	4d	−2, −1, 0, +1, +2	5
	3	4f	−3, −2, −1, 0, +1, +2, +3	7

Electron Configuration

Electron configurations describe the exact arrangement of electrons (given as a superscript) in successive energy levels or shells (1, 2, 3, etc.) and orbitals (s, p, d, f) of an atom, starting with the innermost electrons. For example, a lithium atom's configuration is $1s^22s^1$. The superscripts mean two electrons are in the 1s orbital and one electron is in the 2s orbital. Several "rules" are applied to the filling of electrons:

- **Pauli exclusion principle:** The Pauli exclusion principle states that an orbital can hold a maximum of two electrons if they are of opposite spins. In other words, every electron has a unique set of quantum numbers.
- **Hund's rule:** Hund's rule states that the most stable arrangement of electrons in the same energy level in which electrons have parallel spins (same orientation).
- **Aufbau principle:** The Aufbau principle states that electrons are placed in the most stable orbital. *Aufbau* means "building up" in German (e.g., $1s^2\,2s^2\,2p^6$).

So how is an electron configuration written? First, the number of *total* electrons must be determined. This is equal to the mass number for neutral atoms. For ions, the total electron is corrected for the charge (add electrons for anions; subtract electrons for cations). The electrons are "added" according to Hund's rule and the Aufbau principle. Keep in mind the maximum number of electrons in each type of orbital: s orbitals hold two electrons, p orbitals hold six electrons, d orbitals hold ten electrons, and f orbitals hold 14 electrons.

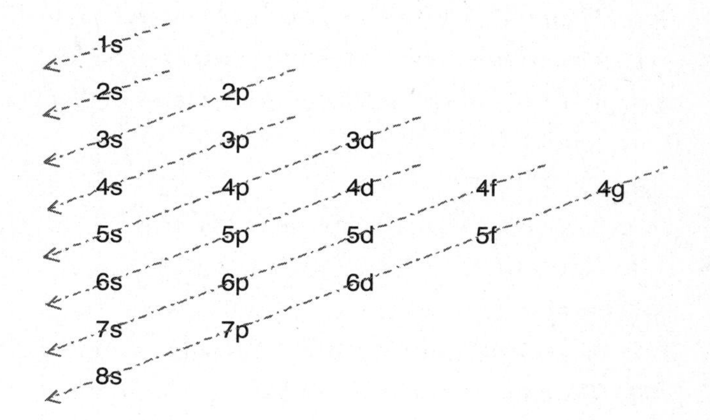

Figure 15.2 Electron-filling sequence

1s → 2s → 2p → 3s → 3p → 4s → 3d → 4p → 5s → 4d → 5p → 6s → 4f → 5d → 6p → 7s → 5d → 6p . . .

Example:
Give the electron configuration for N, Ta, O^{2-}, and Na^+.

N: Total electrons = 7; $1s^22s^22p^3$

Ta: Total electrons = 73;
$1s^22s^22p^63s^23p^64s^23d^{10}4p^65s^24d^{10}5p^66s^24f^{14}5d^3$

O^{2-}: Total electrons = 10 (8 + $2e^-$); $1s^22s^22p^6$

Na^+: Total electrons = 10 (11 − $1e^-$); $1s^22s^22p^6$

Notice that O^{2-} and Na^+ have the same number of electrons. These ions are *isoelectronic* (the same number of electrons).

Because reactions occur primarily with the valance electrons (electrons added since the last noble gas), the electron configurations can be rewritten as a condensed electron configuration identifying the last noble gas in brackets and listing only the valence electrons (see Table 15.2).

Figure 15.2 shows the electron configurations (valance electrons only) for all the elements. Several exceptions are seen and can be attributed to two reasons:

1. Electrons are more stable in full or half-full subshells. Notable examples are Cr, Mo, Cu, Ag, and Au where electrons orient themselves to maximize full and half-full subshells.
2. The d orbitals are more stable than the f orbitals. Notable examples are La, Ac, Ce, and Th.

Practice

Without consulting Figure 15.2, write the electron configurations for

6. Cd
7. Br^-
8. Y
9. V^{5+}
10. P

Table 15.2 Electron Configurations

ATOM OR ION	ELECTRON CONFIGURATION	CONDENSED ELECTRON CONFIGURATION
N	$1s^22s^22p^3$	$[He]2s^22p^3$
Ta	$1s^22s^22p^63s^23p^64s^23d^{10}4p^65s^24d^{10}5p^66s^24f^{14}5d^3$	$[Xe]6s^24f^{14}5d^3$
O^{2-}	$1s^22s^22p^6$	$[He]2s^22p^6$ or [Ne]
Na^+	$1s^22s^22p^6$	$[He]2s^22p^6$ or [Ne]

Orbital Diagrams

Now that you understand electron configurations, an orbital diagram can be drawn. Orbital diagrams represent the orbital where each electron is located. An arrow is used to represent each electron spinning in a particular direction. Recall that s subshells have one orbital, p subshells have three orbitals, and d subshells have five orbitals. The orbital diagram can be constructed using the electron configuration and understanding that Hund's rule states that empty orbitals of a subshell are filled with electrons first before electrons are paired inside orbitals.

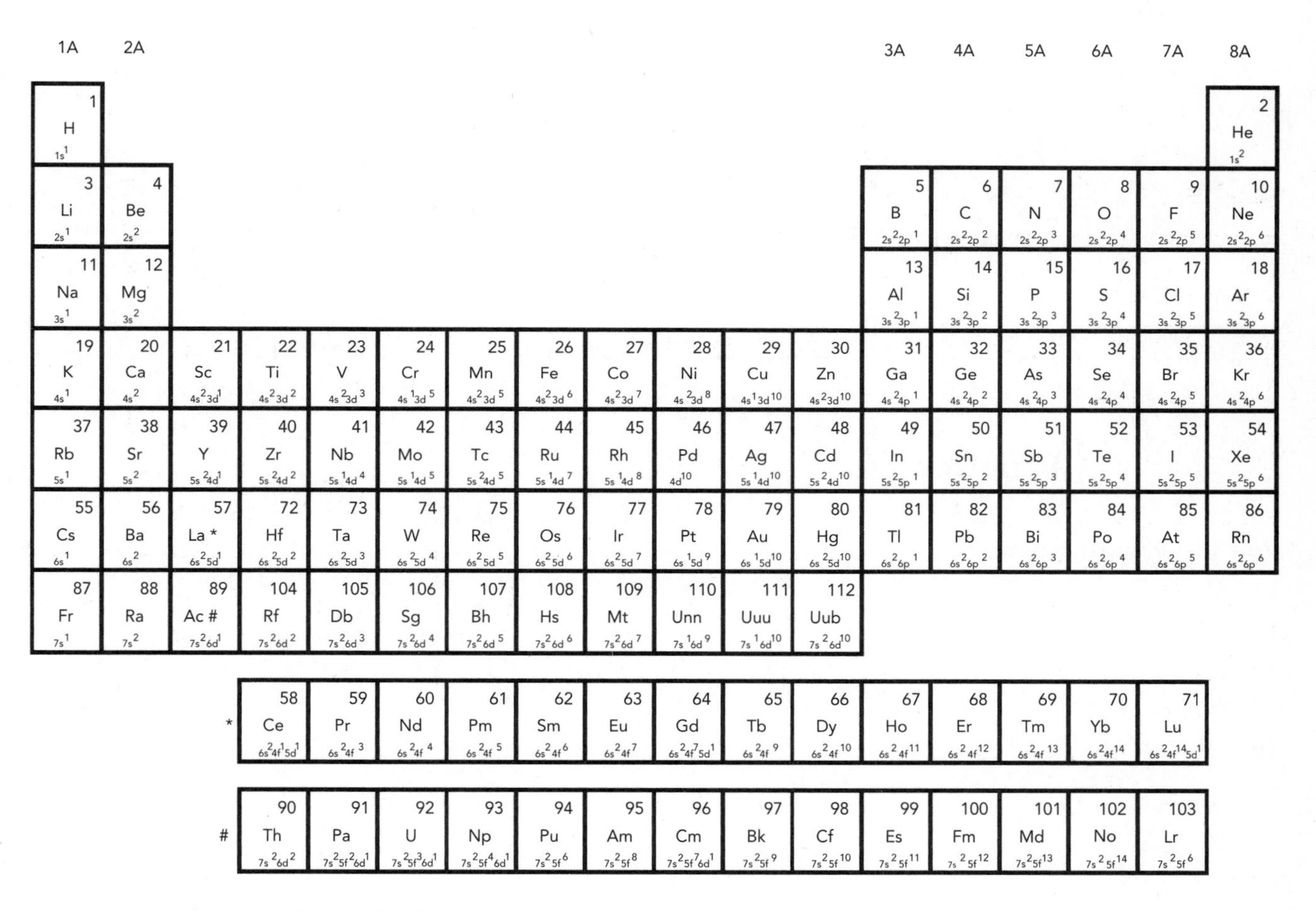

Figure 15.3 Periodic table with electron configurations

Example:

Write the orbital diagram for Li, P, O^{2-}, and Fe.

ATOM OR ION	CONDENSED ELECTRON CONFIGURATION	ORBITAL DIAGRAM
Li	$1s^2 2s^1$	1s: ↑↓ 2s: ↑
P	$[He]2s^2 2p^3$	3s: ↑↓ 3p: ↑ ↑ ↑
O^{2-}	$[He]2s^2 2p^6$	2s: ↑↓ 2p: ↑↓ ↑↓ ↑↓
Fe	$[Ar]4s^2 3d^6$	4s: ↑↓ 3d: ↑↓ ↑ ↑ ↑ ↑

Practice

Write the orbital diagrams for

11. Si

12. Pd

13. N^{3-}

14. B

15. V^{5+}

LESSON 16

Molecular Structure

Three-dimensional structures showing shape, geometry, and valance electrons provide a model for comprehending the arrangement of atoms in a molecule. Molecular geometries play an important part in the intramolecular and intermolecular properties of a substance.

Lewis Structures

Lewis structures are formulas for compounds in which each atom exhibits an octet (eight) of valence electrons. These representations are named after Gilbert N. Lewis for his discovery that atoms in a stable molecule want to achieve a noble gas configuration of eight valance electrons. These electrons are always paired and are represented as dots for nonbonded (lone) pairs or a line for every bonded (shared) pair of electrons. The rules for writing Lewis structures are as follows:

- Sum of all the valance electrons (which should be an even number). Remember: The number of valance electrons is the group number of the element.
- Form bonds between the atoms using pairs of electrons. *Usually*, the least electronegative element is the central element. Hydrogen is never the central element.
- Arrange the remaining electrons as lone pairs or create double or triple bonds to satisfy the octet rule. Exceptions: Hydrogen satisfies the duet (two) rule, and boron and aluminum satisfy the six-electron rule.

Example:

Write the Lewis structure for H_2O, PCl_3, BF_3, and CO_2.

H_2O: Valance electrons: 2(1) + 6 = 8

Form bonds:

H—O—H

Satisfy octet:

H—Ö—H (with two lone pairs on O)

PCl_3: Valance electrons: 5 + 3(7) = 26

Form bonds:

Cl—P(—Cl)—Cl

Satisfy octet:

:Cl: —P— :Cl: (each Cl with three lone pairs, one lone pair on P)

BF_3: Valance electrons: 5 + 3(7) = 26

Form bonds:

F—B(—F)—F

Satisfy octet:

:F: —B— :F: (each F with three lone pairs)

(Boron follows the six-electron rule.)

CO_2: Valance electrons: 4 + 2(6) = 16

Form bonds:

O—C—O

Satisfy octet:

:Ö—C—Ö: (each O with three lone pairs)

In trying to satisfy the octet and only have 16 valence electrons, the carbon was not fulfilled with only single bonds.

:O=C=O: (each O with two lone pairs)

A pair of electrons from each oxygen was used to form a double bond to the carbon and satisfy the octet of all three atoms.

Another way to test to see if the octet rule is met is to write all the paired electrons as dots and circle each element. Each element circled should have eight electrons (two for hydrogen and six for boron and aluminum) and clearly show the bond overlap for the bonding pairs.

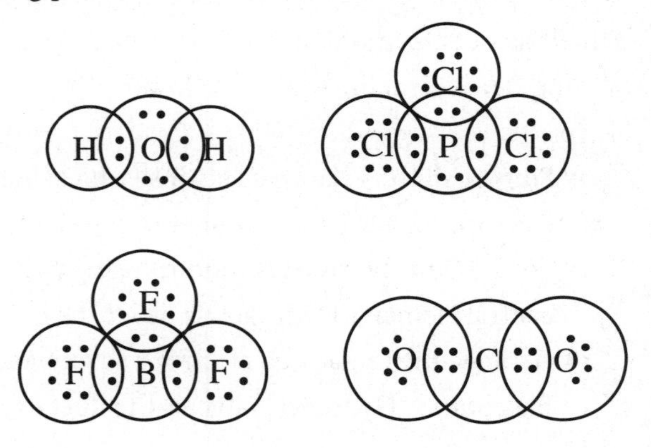

Practice

Write the Lewis structures for the following molecules:

1. SiH_4
2. $AlCl_3$
3. HCN
4. NH_3
5. $HClO_3$ (Hint: Acidic hydrogens are attached to a oxygen.)
6. H_2CO (Hint: There are no acidic hydrogens.)
7. XeF_4

Molecular Geometry: Valence Shell Electron Pair Repulsion (VSEPR) Theory

The *VSEPR model* is based on electrostatic repulsion among electron pair orbitals. By pushing each pair as far as possible, electron pairs dictate which geometry or shape a molecule will adopt. Molecules should be written as 2-D Lewis structures, and then determine the number of bonding pairs and nonbonding pairs. A summary of the shapes and possible arrangements can be found in Figure 16.1 and Table 16.1. Double and triple bonds can be treated as one bonding pair for VSEPR theory.

Table 16.1 Possible Arrangement of Electrons about a Central Atom

MOLECULE CLASS	TOTAL ELECTRON PAIRS	BONDING PAIRS	NON-BONDING PAIRS	ARRANGEMENT OF ELECTRONS	GEOMETRY (OR SHAPE)	EXAMPLE
AB_2	2	2	0	Linear	Linear	$BeCl_2$, CO_2
AB_3	3	3	0	Trigonal planar	Trigonal planar	BH_3
AB_4	4	4	0	Tetrahedral	Tetrahedral	CH_4
AB_3E	4	3	1	Tetrahedral	Trigonal pyramidal	NH_3
AB_2E_2	4	2	2	Tetrahedral	Bent (angular or V-shaped	H_2O
AB_5	5	5	0	Trigonal bipyramidal	Trigonal bipyramidal	PCl_5
AB_4E	5	4	1	Trigonal bipyramidal	Seesaw	SF_4
AB_3E_2	5	3	2	Trigonal bipyramidal	T-shaped	ClF_3
AB_2E_3	5	2	3	Trigonal bipyramidal	Linear	XeF_2
AB_6	6	6	0	Octahedral	Octahedral	SF_6
AB_5E	6	5	1	Octahedral	Square pyramidal	BrF_5
AB_4E_2	6	4	2	Octahedral	Square planar	XeF_4

A = central atom, B = bonding atom, E = nonbonding electron pair

Example:
Using VSEPR, predict the shape for the following molecules or ions: KrF_2, HCN, PCl_3, NO_2^-, NO_3^-.

Molecule or ion	Lewis structure	Bonding electron pairs	Nonbonding electron pairs	Shape
KrF_2	F—Kr—F	2	3	Linear
HCN	H—C≡N:	2*	0	Linear
PCl_3	:Cl—P—Cl: with :Cl: above P	3	1	Trigonal pyramidal
NO_2^-	:O—N=O:	2*	1	Bent
NO_3^-	:O—N—O: with :O: double-bonded above N	3*	0	Trigonal planar

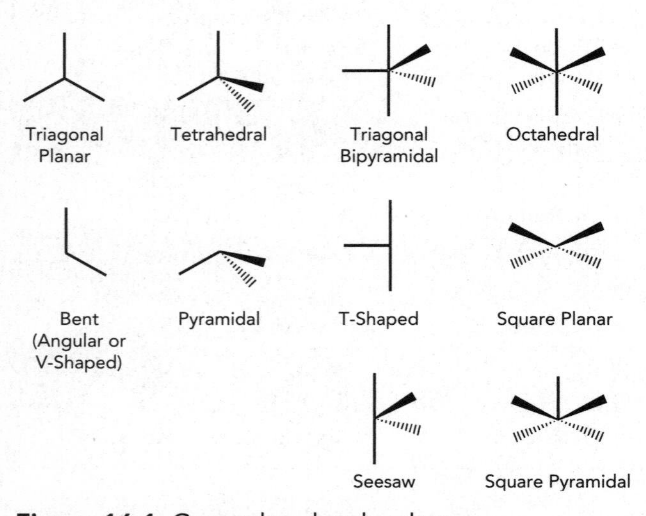

Figure 16.1 General molecular shapes

Practice

Using VSEPR, predict the shape for the following molecules or ions.

8. $AlCl_3$
9. O_3
10. H_2CO
11. SnH_4
12. $XeOF_4$
13. SF_4

Resonance Structures

Resonance occurs when one or more valid Lewis structures exist for a molecule or polyatomic ion. The structures that represent the substance are called resonance structures. Each resonance structure does not characterize the substance, but the average of all the resonance structures represents the molecule or polyatomic ion. Resonance structures are usually placed in brackets and separated by a double-headed arrow (↔).

Example:

Show the resonance structures for nitrate, NO_3^-. Although the three nitrate resonance structures are written separately, nitrate is a combination of all three structures.

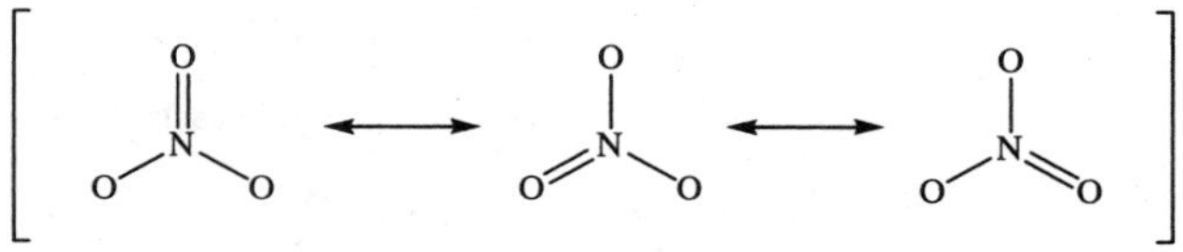

Practice

Show the resonance structures for the following molecules or polyatomic ions.

14. O_3
15. CO_3^{2-}
16. HNO_3
17. SO_3^{2-}

Formal Charge

Atoms in certain molecules or polyatomic ions may have a formal charge. A *formal charge* is the difference in the number of valance electrons in the neutral atom (group number) and the number of electrons assigned to that atom in the molecule or polyatomic ion. Mathematically, the equation is

formal charge = group number − number of bonds − number of lone electrons

Example:

Give the formal charge for each atom in the nitrate ion, NO_3^-, and ozone, O_3 (yes, even neutral molecules can have elements with formal charges).

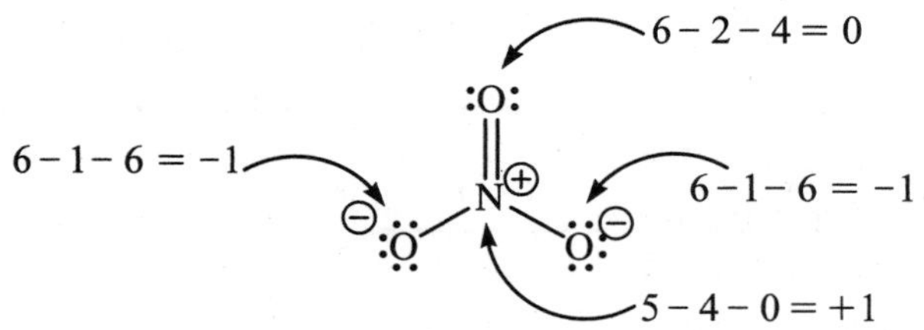

$6-2-4=0$

$6-1-6=-1$

$6-1-6=-1$

$5-4-0=+1$

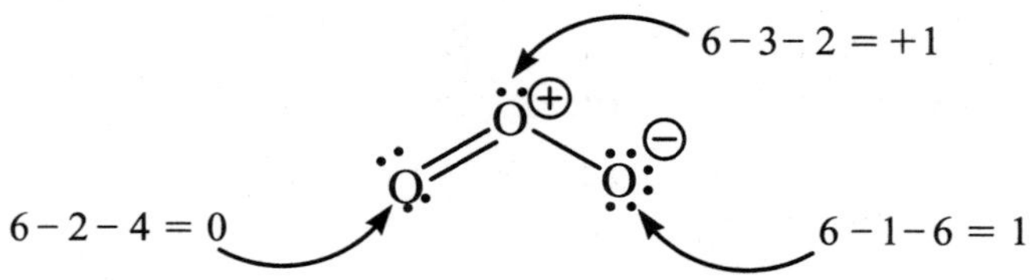

$6-3-2=+1$

$6-2-4=0$

$6-1-6=1$

Note

All formal charges must add up to the charge on the polyatomic ion or zero for a neutral molecule.

Practice

Give the formal charge for each atom in the following molecules or polyatomic ions.

18. SO_4^{2-}
19. CO_3^{2-}
20. HNO_3
21. SO_3^{2-}

Polarity

The *electronegativity* of an element is its strength and ability to attract paired electrons in a covalent molecule. Electronegativity increases as you move up and to the right on the periodic table. Fluorine is the most electronegative element. (See Figure 16.2)

A *dipole* results in a covalent bond between two atoms of different electronegativity. A partial positive (δ+) and a negative charge (δ−) develop at both ends of the bond, creating a dipole (i.e., two poles) oriented from the positive end to the negative end. The oxygen atom is more electronegative than hydrogen in water and the result is a dipole. Dipoles are represented by a line with a perpendicular line (⟼), the arrow points the atom with greater negative charge.

A dipole moment will exist in a molecule if the resulting dipoles do not cancel based on their additive vectors. If the two dipoles of water are added, water has a dipole moment in the "up" direction.

H 2.1																
Li 1.0	Be 1.5											B 2.0	C 2.5	N 3.0	O 3.5	F 4.0
Na 0.9	Mg 1.2											Al 1.5	Si 1.8	P 2.1	S 2.5	Cl 3.0
K 0.8	Ca 1.0	Sc 1.3	Ti 1.5	V 1.6	Cr 1.6	Mn 1.5	Fe 1.8	Co 1.9	Ni 1.9	Cu 1.9	Zn 1.6	Ga 1.6	Ge 1.8	As 2.0	Se 2.4	Br 2.8
Rb 0.8	Sr 1.0	Y 1.2	Zr 1.4	Nb 1.6	Mo 1.8	Tc 1.9	Ru 2.2	Rh 2.2	Pd 2.2	Ag 1.9	Cd 1.7	In 1.7	Sn 1.8	Sb 1.9	Te 2.1	I 2.5
Cs 0.7	Ba 0.9	La 1.0	Hf 1.3	Ta 1.5	W 1.7	Re 1.9	Os 2.2	Ir 2.2	Pt 2.2	Au 2.4	Hg 1.9	Tl 1.8	Pb 1.9	Bi 1.9	Po 2.0	At 2.2
Fr 0.7	Ra 0.9															

Figure 16.2 Electronegativities

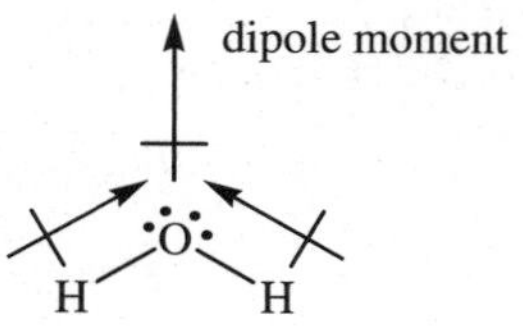

dipole moment

H—Ö—H (water, with bond dipoles and net dipole moment arrow)

Example:
Identify the dipole for each of the following bonds: B-F, Cl-I, N-H.

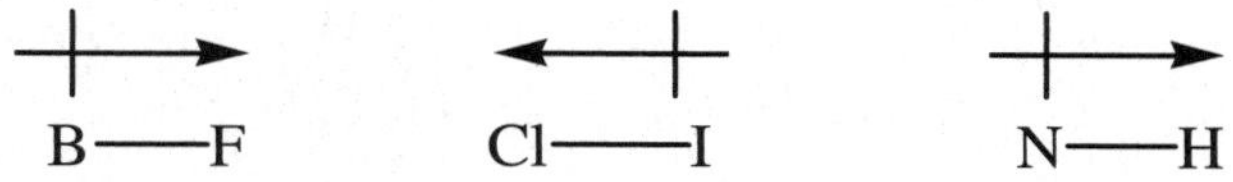

B—F (arrow toward F)   Cl—I (arrow toward Cl)   N—H (arrow toward H)

Predict whether each of the following molecules has a dipole moment: SCl_2, CH_2Cl_2.
SCl_2: Sulfur dichloride has a bent structure and a dipole:

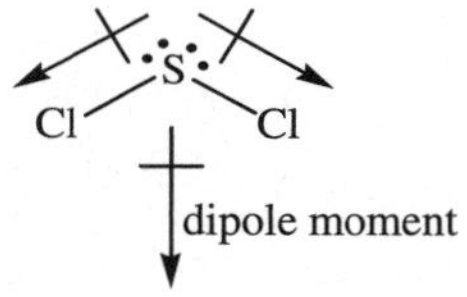

Cl—S—Cl

dipole moment

CH_2Cl_2: If drawn (incorrectly!) planar, the molecule shows no dipole:

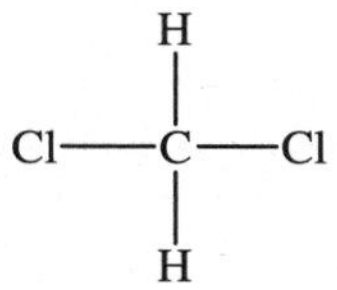

H

Cl—C—Cl

H

However, if drawn correct as tetrahedral, the dipole is:

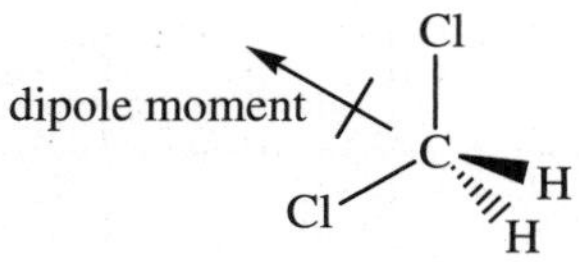

Cl

dipole moment

C H H

Cl

Practice

Identify the dipole for each of the following bonds:

22. H-Si
23. C-F
24. B-Cl
25. S-F

Predict whether each of the following molecules has a dipole moment:

26. PCl_3
27. BF_3
28. CH_3Cl

LESSON

17 Molecular Orbital Theory

In the previous lesson, we used Lewis structures to describe how atoms form bonds. *Molecular orbital theory* provides an alternate and deeper description of molecules. In many cases, the two models complement each other; however, molecular orbital theory is able to explain certain observed properties that simpler models cannot.

Combinations of Atomic Orbitals

You should be familiar with the concept of atomic orbitals from previous lessons. These orbitals are defined by the quantum numbers n, l, and m_l. where n is the principal quantum number, l is the orbital quantum number, and m_l is the magnetic quantum number. Orbitals describe the shape and size of regions where an electron (or pair of electrons) is likely to be found. These numbers are often translated to more familiar notation (e.g. 1s, 2s, $2p_x$, $2p_y$, $2p_z$).

How can we use our knowledge of the shapes of atomic orbitals to understand bonding? Molecular orbital theory uses combinations of atomic orbitals to create bonding and antibonding molecular orbitals. When two atomic orbitals are brought closer together, they begin to interact and form a molecular orbital. Atomic and molecular orbitals exist in two phases (positive and negative) and this affects the way they interact when brought together. If two orbitals are in the same phase, the overlap is constructive and a bonding molecular orbital is formed. If the two orbitals are in opposite phases, the overlap is destructive and an antibonding molecular orbital is created. Bonding orbitals have shapes that place electron density between the two atoms, whereas antibonding orbitals place electron density away from the center.

Now that we have discussed the basic idea behind constructing molecular orbitals, we will put it to good use. Let's look first at how two 1s orbitals combine (Fig 17.1). When two orbitals overlap in the same phase, electron density builds in between the two atomic nuclei. Conversely, when they overlap out of phase, electron density in between the two nuclei is cancelled out, creating an antibonding orbital. This type of bonding, where electron density lines up along the atomic axis, is referred to as σ-bonding (the antibonding orbitals are as σ-bonding, or σ*, orbitals).

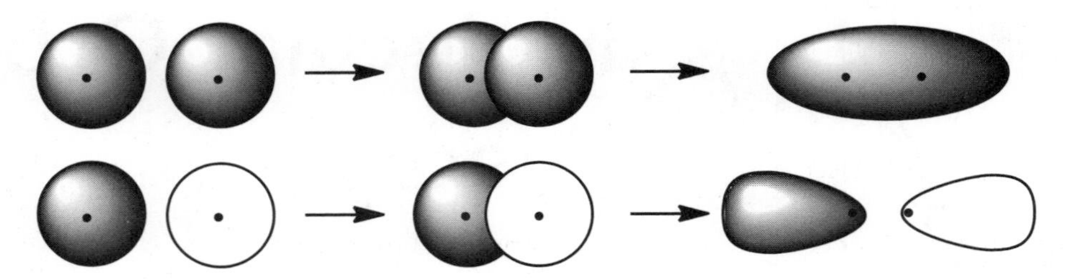

Figure 17.1 Bonding and antibonding molecular orbitals constructed from two 1s atomic orbitals

Bonding molecular orbitals are always lower in energy than the atomic orbitals that created them and antibonding orbitals are always higher in energy. Every electron in a bonding orbital contributes one half of a bond and every electron in an antibonding orbital removes one half of a bond. So, the overall *bond order* of for a molecular bond is:

$$\text{Bond Order} = \frac{(\text{\# bonding electrons} - \text{\# antibonding electrons})}{2}$$

Combinations With p Orbitals

Now that we have explored how two s orbitals interact to form bonding and antibonding orbitals, let's explore how p orbitals are treated. P orbitals can overlap one of two ways: end-on, where two lobes directly face each other, and side-one, where the orbitals point parallel to each other. When p orbitals combine in an end-on fashion, they create σ bonds, much like the s orbitals.

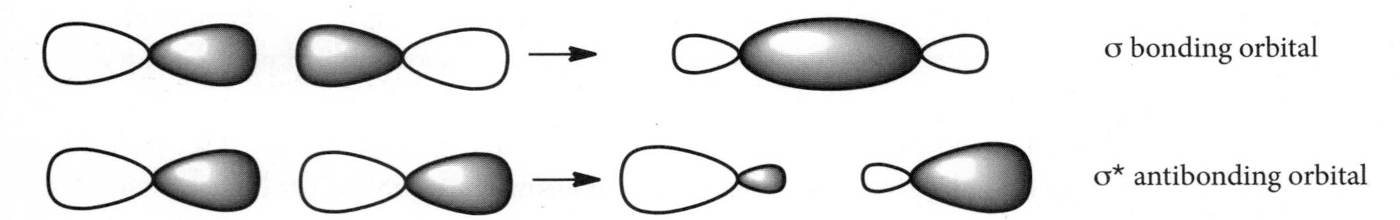

Figure 17.2 σ and σ* molecular orbitals constructed from two p atomic orbitals

However, when they combine side-on, they form a different type of bond, a π bond. These bonds do not have electron density directly between the two nuclei. Instead, the electron density runs just above and below the axis. When p orbitals are mixed side-on such that their lobes line up out-of-phase, a π* (antibonding) molecular orbital is formed.

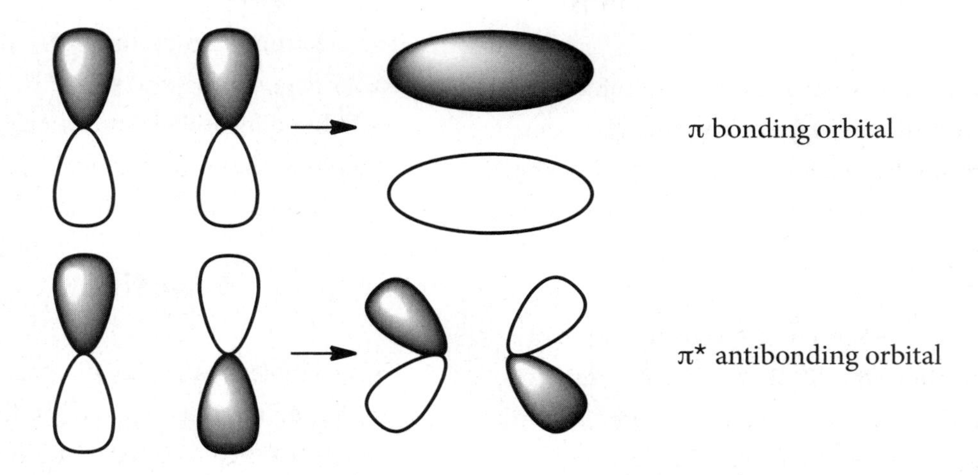

Figure 17.3 π and π* molecular orbitals constructed from two p atomic orbitals

Constructing a Molecular Orbital Diagram

Now that we know how orbitals combine, we can use this information to construct a molecular orbital diagram. We will first examine the simplest molecule, hydrogen (H_2). The 1s orbitals of two hydrogen atoms combine to form a lower energy σ orbital and a higher energy σ* orbital. Each hydrogen atom contributes its one electron to the lower energy σ orbital (since an orbital can hold two electrons).

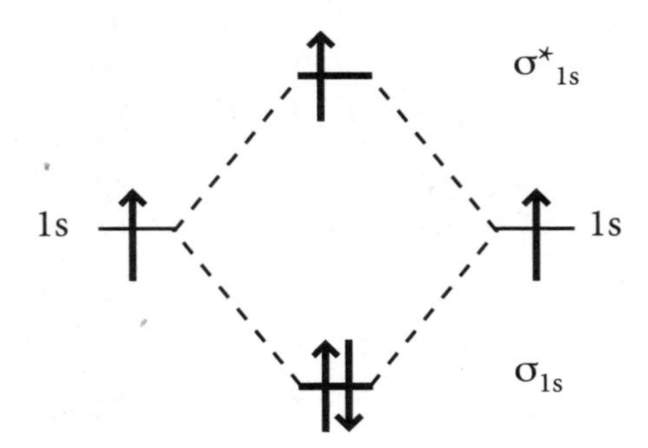

The diagram shows that H_2 has two electrons in bonding orbitals and zero electrons in nonbonding orbitals, so its overall bond order is:

$$\text{H-H bond order} = \frac{(2 \text{ bonding electrons} - 0 \text{ non-bonding electrons})}{2 = 1}$$

This treatment can easily be expanded for more complicated molecules. The atomic orbitals are placed in order of increasing energy (1s, 2s, 2p, etc). The matching atomic orbitals mix together to form the same molecular orbitals discussed above. Remember, bonding orbitals are lower in energy than antibonding orbitals. The electrons from the atomic orbitals are used to fill the new molecular orbitals from lowest to highest energy.

Example

Generate the molecular orbital diagram for molecular nitrogen, N_2. Give the number of bonding electrons, antibonding electrons, and bond order for the molecule.

Solution

The nitrogen atom has seven electrons and an electronic configuration of $1s^2 2s^2 2p^3$. We can construct ten molecular orbitals from the ten atomic orbital (five from each N). We then fill them with 14 electrons.

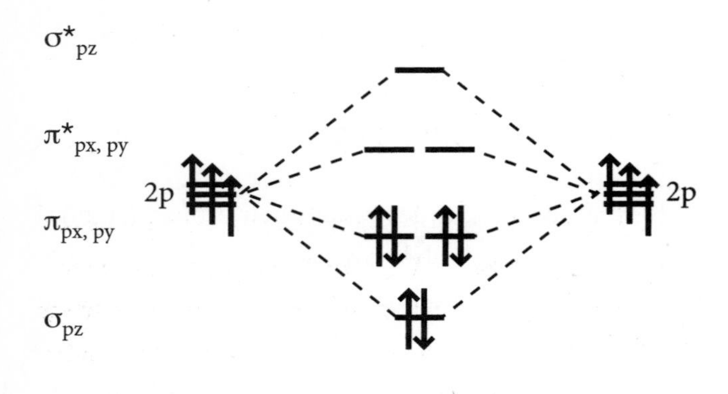

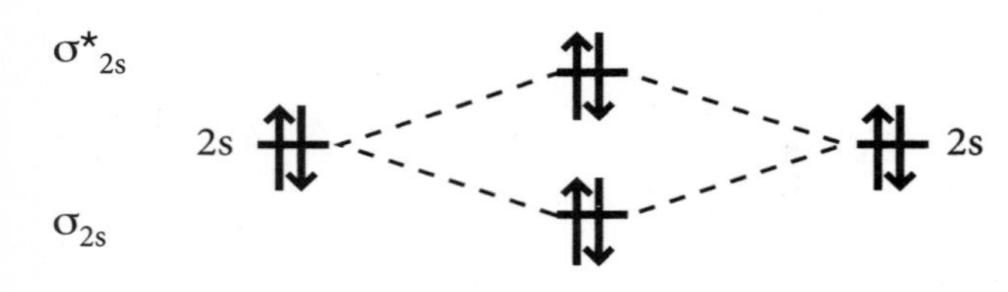

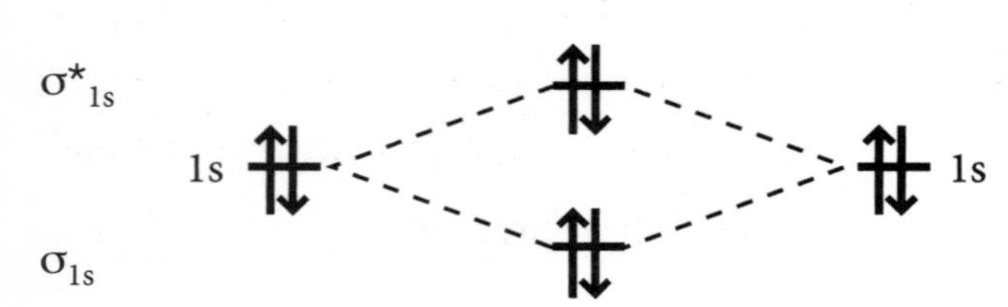

The electronic structure for N_2 is $(\sigma_{1s})^2(\sigma_{1s}{}^*)^2(\sigma_{2s})^2(\sigma_{2s}{}^*)^2(\sigma_{2p})^2(\pi_{2p})^4(\pi_{2p}{}^*)^0(\sigma_{2p}{}^*)^0$. There are ten electrons in bonding orbitals and four electrons in antibonding orbitals. The bond order for N_2 is $(10 - 4)/2 = 3$.

Practice

1. Using a molecular orbital diagram, give the number of bonding and antibonding electrons and predict the bond order of Be_2. Does Be_2 likely exist in nature?

2. Using a molecular orbital diagram, give the number of bonding and antibonding electrons and predict the bond order of F_2?

3. Construct a molecular orbital diagram for $He_2{}^+$. What is its bond order?

4. Using a molecular orbital diagram, give the number of bonding and antibonding electrons and predict the bond order of Li_2. Does Li_2 likely exist in nature?

5. A diatomic molecule contains six electrons in bonding orbitals and two electrons in antibonding orbitals. What is the bond order?

LESSON

18 Magnetism

We all know the feeling when a proud parent uses a magnet to put the latest report card up on the refrigerator. Magnets are everywhere, but how often have you stopped to consider the properties of magnets and magnetism? Chemists study the magnetic properties of molecules to gain clues about their electronic structure.

Three Types of Magnetism

Magnetism describes the way a material or compound responds to being placed in a magnetic field. To the non-scientist, metals such as iron are thought of as being the only compounds with magnetic properties, but all compounds exhibit some form of magnetism. Chemists define the type of magnetism of most molecules as falling in one of three categories: diamagnetism, paramagnetism, and ferromagnetism.

Diamagnetism

Diamagnetic compounds are repelled by an external magnetic field. This is a relatively weak form of magnetism and is commonly associated with materials such as water, copper, and even diamonds that a non-chemist would describe as nonmagnetic. In terms of their electronic structure, diamagnetic molecules have an electronic structure in which all electrons are paired. How does this relate to the molecular orbital diagrams we developed earlier? Let's take a look at the simplest diamagnet: H_2. As you can see, there is one unpaired electron in the 1s orbital on each hydrogen atom which pair to form a σ-bonding molecular orbital. There are no unpaired electrons, and H_2 is a diamagnetic molecule.

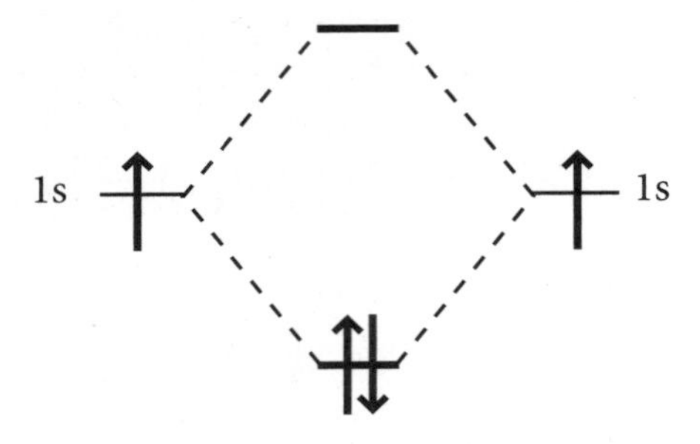

Figure 18.1 Molecular orbital diagram for dihydrogen

Paramagnetism

Paramagnetic compounds are attracted to an external magnetic field. That is, when placed next to a magnet, the compound will be pulled towards the magnet. A chemist interprets *paramagnetism* to mean that the molecule in question has one or more unpaired electrons. How do unpaired electrons relate to a molecule's response to a magnetic field? Electrons possess an intrinsic charge (−1) and spin (up or down). These properties create a magnetic dipole, which results in the creation of a magnetic field. In the absence of an external magnetic field, these dipoles are disordered and point in all directions. The magnetism of each molecule is cancelled out by all the molecules. It is only when placed in a magnetic field that these molecular dipoles align (pointing towards the magnet) and the substance exhibits the properties of a magnet.

How do chemists use this information to learn something useful about the molecules being studied? Molecular oxygen (O_2) provides a great example. An oxygen atom has the electronic structure $1s^2 2s^2 2p^4$, and there are two ways we can construct a molecular orbital diagram, depending on the ordering of the σ-bonds or π-bonds:

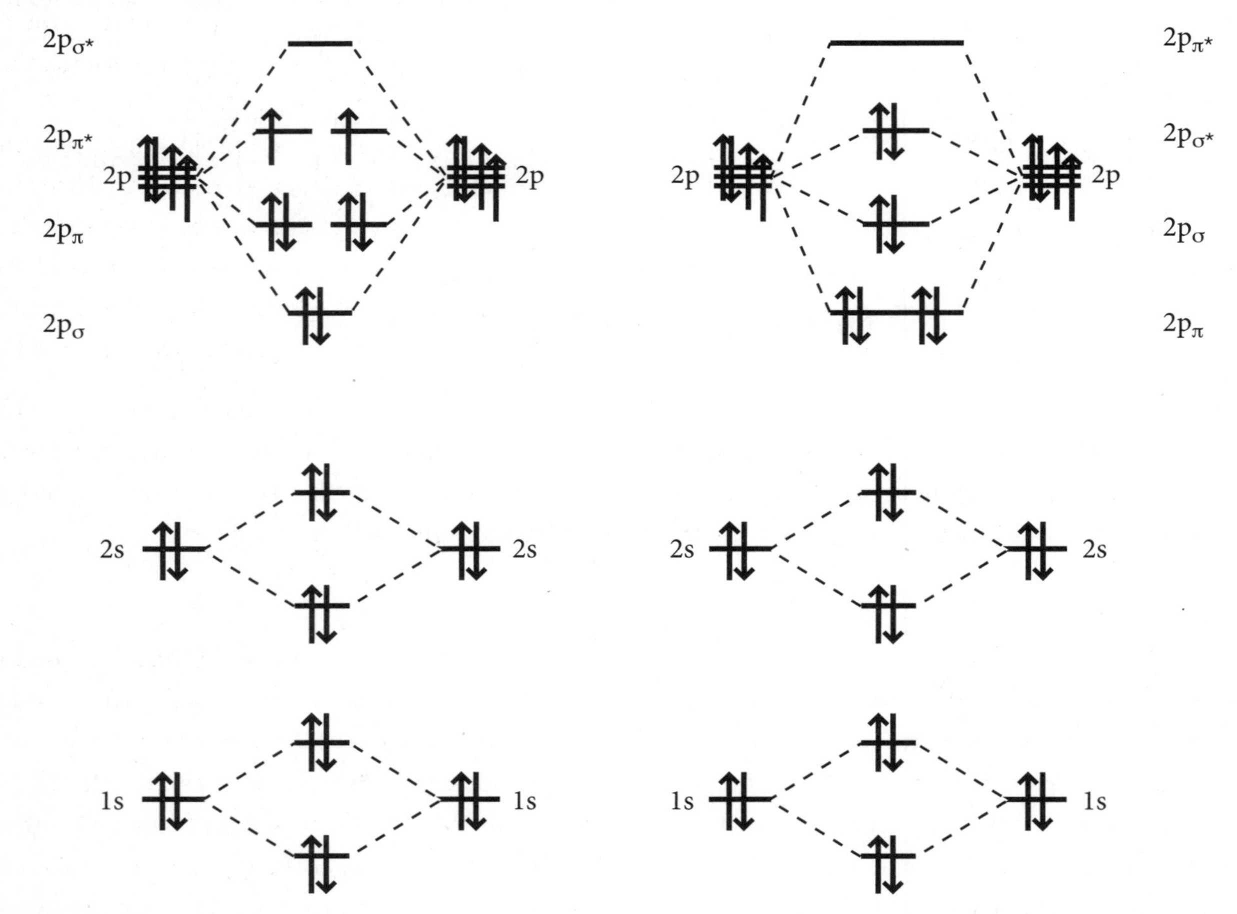

Figure 18.2 Two potential molecular orbital diagrams for dioxygen. The first one is correct.

Both molecular orbital diagrams seem reasonable, so how do chemists determine which is correct? The magnetic properties of oxygen provide a clue. If the π* orbitals are filled before the σ* orbitals, then there are two unpaired electrons and O_2 is paramagnetic. If the σ* orbitals are filled before the π*orbitals, then all electrons are paired and O_2 is diamagnetic. When liquid oxygen is poured between the poles of a magnet, it will concentrate between the poles and float there until it evaporates. The magnetic properties of oxygen show that the highest occupied molecular orbital is 2pπ* rather than 2pσ*.

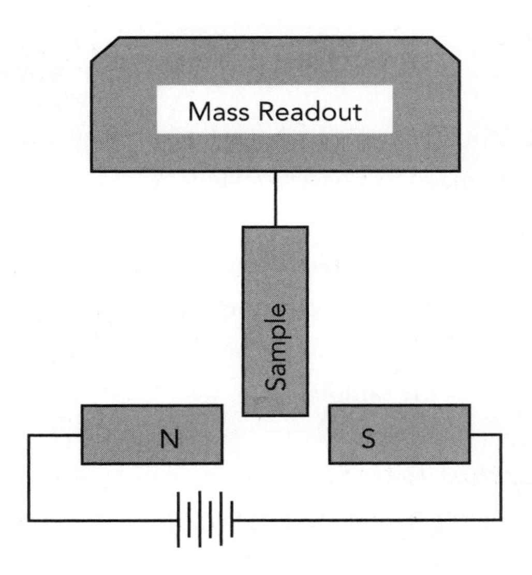

Figure 18.3 A Gouy Balance

Gouy balance

We now know that the magnetic properties of a substance can provide critical information about its electronic structure. But how do chemists determine whether a compound is diamagnetic or paramagnetic? Scientists use a Gouy balance to take advantage of the different ways diamagnetic and paramagnetic materials respond to an external magnetic field. A *Gouy balance* is similar to a typical balance, except that it has been combined with an electromagnet. Here is how it works: A sample is placed on the balance and the mass is recorded. Then, the electromagnet is activated and the mass is collected again. If the compound is paramagnetic, it is attracted to the magnet and the apparent mass increases. If it is diamagnetic, it is repelled by the magnet and its apparent mass decreases.

Ferromagnetism

Ferromagnetism is the type of magnetism one typically associates with a magnet. It is the strongest type of magnetism. In fact, it is the only type of material that can generate magnetic fields strong enough to be easily felt. Like paramagnets, ferromagnets are composed of atoms with unpaired electrons. However, while paramagnets require an external magnetic field to align the individual dipoles within the material, the dipoles within ferromagnets are able to align on their own. Thus, they are able to act as magnets even in the absence of an external magnetic field.

Practice

1. Would you expect N_2 to be paramagnetic, diamagnetic, or ferromagnetic?

2. A 542 mg sample is placed on a Gouy balance. When the electromagnetic is activated, the mass reads 522 mg.What can you conclude about the sample?

3. Which type of magnetism would you expect a hydrogen *atom* to display?

4. Which of the following atoms would you expect to be paramagnetic: He, Li, Be, B?

5. A refrigerator magnet is an example of what kind of magnetism?

LESSON

19 Oxidation-Reduction

Most aqueous reaction equations can be balanced by trial and error. *Oxidation-reduction* reactions require a more systematic approach to balancing equations.

Balancing Redox Reactions

Redox reactions involve the transfer of electrons from one metal to another. The oxidation and reduction parts of the reaction can be broken down into their half reactions. The *half-reaction* contains only the compounds that contain the species being reduced *or* the species being oxidized. As an example, copper metal reacts with a silver ion to yield silver metal and copper ion:

$$Cu(s) + 2Ag^{+}(aq) \rightarrow 2Ag(s) + Cu^{2+}(aq)$$

The oxidation half-reaction involves the oxidation of copper:

$$Cu(s) \rightarrow Cu^{2+}(aq) + 2e^{-}$$

The reduction half-reaction involves the reduction of the silver ions:

$$Ag^{+}(aq) + e^{-} \rightarrow Ag(s)$$

Balancing oxidation-reduction reactions depends on whether the solution is acidic or basic.

The method for balancing redox reactions in an acidic solution is as follows:

1. Write the reduction and oxidation half-reactions. For each half-reaction:
   - **a.** Balance all the elements except oxygen and hydrogen.
   - **b.** Balance the oxygen using H_2O.
   - **c.** Balance the hydrogens using H^+.
   - **d.** Balance the charge using electrons (e^-).
2. If the number of electrons needed previously in 1d for each half-reaction is different, equalize the electrons in each half-reaction by multiplying the appropriate reaction(s) by an integer.
3. Add the half-reactions and check that the elements and charges are balanced.

Example:
Balance the following reaction in an acidic solution:

$$H_2SO_4 + C \rightarrow CO_2 + SO_2$$

Solution:
Sulfur is being reduced (S^{6+} to S^{4+}) and carbon is being oxidized (C^0 to C^{4+}).

1. Write the reduction and oxidation half-reactions:

   Oxidation: $C \rightarrow CO_2$

   Reduction: $H_2SO_4 \rightarrow SO_2$

   For each half-reaction:

   a. Balance all the elements except oxygen and hydrogen:

   Oxidation: $C \rightarrow CO_2$

   Reduction: $H_2SO_4 \rightarrow SO_2$

   b. Balance the oxygen using H_2O:

   Oxidation: $C + 2H_2O \rightarrow CO_2$

   Reduction: $H_2SO_4 \rightarrow SO_2 + 2H_2O$

   c. Balance the hydrogens using H^+:

   Oxidation: $C + 2H_2O \rightarrow CO_2 + 4H^+$

   Reduction: $2H^+ + H_2SO_4 \rightarrow SO_2 + 2H_2O$

   d. Balance the charge using electrons (e^-):

   Oxidation: $C + 2H_2O \rightarrow CO_2 + 4H^+ + 4e^-$

   Reduction: $2e^- + 2H^+ + H_2SO_4 \rightarrow SO_2 + 2H_2O$

2. If the number of electrons needed in 1 d for each half-reaction is different, equalize the electrons in each half-reaction by multiplying the appropriate reaction(s) by an integer. In this case, the reduction half-reaction can be multiplied by 2 to yield both half-reactions with four electrons. Notice that the electrons must be on opposite sides to cancel.

   Oxidation: $C + 2H_2O \rightarrow CO_2 + 4H^+ + 4e^-$

   Reduction: $4e^- + 4H^+ + 2H_2SO_4 \rightarrow 2SO_2 + 4H_2O$

3. Add the half-reactions and check that the elements and charges are balanced.

   $C + \cancel{2H_2O} \rightarrow CO_2 + \cancel{4H^+} + \cancel{4e^-}$

   $\cancel{4e^-} + \cancel{4H^+} + 2H_2SO_4 \rightarrow 2SO_2 + \cancel{4H_2O}2H_2O$

   $2H_2SO_4 + C \rightarrow CO_2 + 2SO_2 + 2H_2O$

Practice

Balance the following reactions in acidic solutions:

1. $Cu + NO^{3-} \rightarrow Cu^{2+} + NO$
2. $MnO_4^- + Fe^{2+} \rightarrow Fe^{3+} + Mn^{2+}$
3. $Zn + HCl \rightarrow Zn^{2+} + H_2 + Cl^-$

The method for balancing redox reactions in a basic solution is as follows:

1. Balance the equations as if an acidic solution.
2. Add an equal number of OH^- ions to each side of the reaction, cancel out the H^+ ions, and form water on the side containing both H^+ and OH^-.
3. If necessary, cancel out water (H_2O) and check that the elements and charges are balanced.

Example:
Balance the following reaction in basic solution:
$NH_3 + ClO^- \rightarrow Cl_2 + N_2H_2$

Solution:

1. Balance the equations as if an acidic solution.

   $2H^+ + 2NH_3 + 2ClO^- \rightarrow$

   $Cl_2 + N_2H_4 + 2H_2O$

2. Add an equal number of OH^- ions to each side of the reaction, cancel out the H^+ ions, and form water on the side containing both H^+ and OH^-.

   $2OH^- + 2H^+ + 2NH_3 + 2ClO^- \rightarrow$
   $Cl_2 + N_2H_4 + 2H_2O + 2OH^-$

   $2H_2O + 2NH_3 + 2ClO^- \rightarrow Cl_2 + N_2H_4 +$
   $2H_2O + 2OH^-$

3. If necessary, cancel out water (H_2O) and check that the elements and charges are balanced.

   $2NH_3 + 2ClO^- \rightarrow Cl_2 + N_2H_4 + 2OH^-$

Practice

Balance the following reactions in basic solutions:

4. $Cl_2 \rightarrow ClO^- + Cl^-$
5. $H_2O_2 + SO_3^{2-} \rightarrow H_2O + SO_4^{2-}$
6. $Fe^{2+} + H_2O_2 \rightarrow Fe^{3+}$

LESSON

20 Organic Chemistry

Organic chemistry is the study of the structure and reactivity of compounds containing carbon. Carbon is the backbone of more compounds than all other elements combined. The versatility of carbon allows for an infinite number of possible compounds and infinite possibilities for the discovery of new and exciting drugs, plastics, and other innovations.

Introduction

As stated, organic chemistry is the study of compounds containing carbon. Most organic compounds also contain hydrogen, and many contain heteroatoms such as oxygen, nitrogen, the halogens, phosphorus, and sulfur. Nucleic acids (DNA, RNA, etc.), proteins, carbohydrates (sugars), fats (lipids), plastics, and petroleum products are just a *few* classes of organic compounds.

Organic chemistry can be traced back to the nineteenth century when German chemist Friedrich Wöhler synthesized the first organic compound urea, a component of urine.

$$NH_4^+ NCO^- \xrightarrow{\text{Heat}} H_2N-C(=O)-NH_2$$

Ammonium cyanate
inorganic

Urea
organic

Carbon is *tetravalent* (forming four bonds) and can form single bonds, double bonds, and triple bonds. As seen in Table 20.1, the four types of hydrocarbons are alkanes (single bonds), alkenes (double bonds), alkynes (triple bonds), and aromatic. Aromatics are unsaturated hydrocarbons that have cyclic structures. A common and representative compound for aromatic is benzene.

Nomenclature of Alkanes

Alkanes are organic molecules in which all the carbons are bonded to four atoms (i.e., all single bonds). These molecules are saturated because carbon has the maximum number of atoms surrounding it. Organic molecules are named systematically using a straight-chain or unbranched alkane as a backbone (see Table 20.2).

Table 20.1 Hydrocarbon Properties

HYDROCARBON	SATURATION	BONDING	STRUCTURE	EXAMPLE
Alkane	Saturated	Single bonds	—C—C—	Methane, Ethane
Alkene	Unsaturated	Double bonds	C=C	Ethene, Propene
Alkyne	Unsaturated	Triple bonds	—C≡C—	Ethyne
Aromatic	Unsaturated		benzene "an aromatic compound"	Benzene

Table 20.2 Parent Names of Unbranched Alkanes

PREFIX	NUMBER OF CARBON ATOMS	NAME	STRUCTURE
meth–	1	Methane	CH_4
eth–	2	Ethane	CH_3CH_3
prop–	3	Propane	$CH_3CH_2CH_3$
but–	4	Butane	$CH_3(CH_2)_2CH_3$
pent–	5	Pentane	$CH_3(CH_2)_3CH_3$
hex–	6	Hexane	$CH_3(CH_2)_4CH_3$
hept–	7	Heptane	$CH_3(CH_2)_5CH_3$
oct–	8	Octane	$CH_3(CH_2)_6CH_3$
non–	9	Nonane	$CH_3(CH_2)_7CH_3$
dec–	10	Decane	$CH_3(CH_2)_8CH_3$

Rules for naming alkanes:

- Locate the longest continuous chain of carbons. This chain will determine the backbone or parent name of the molecule.
- Locate the substituents or side groups attached to the longest chain. Use a prefix to identify the number of carbons and an *–yl* suffix to indicate that the group is a substituent. When two or more of the same substituent is present, use the prefixes *mono–*, *di–*, *tri–*, *tetra–*, and so on.
- Number the longest chain to give the substituents a location number. Use the smallest possible numbers.
- Regardless of the location numbers, the substituents are listed alphabetically in the name. A comma separates numbers, and a dash separates numbers and letters.

Example:

Give the IUPAC systematic name for the following molecules:

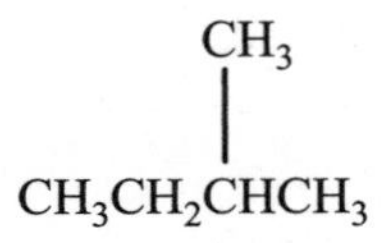

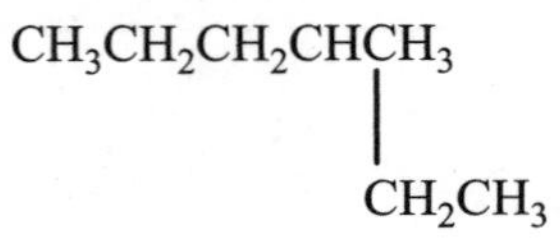

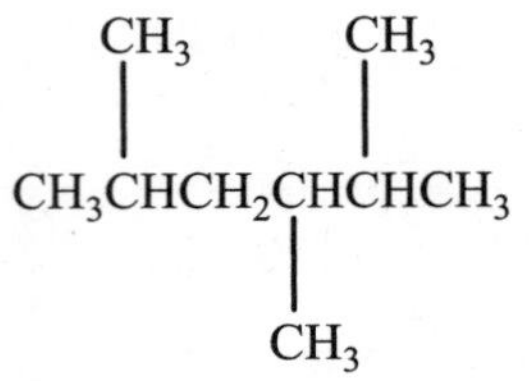

$$CH_3CH_2CHCH_3 \text{ (with } CH_3 \text{ on C-3)}$$

$$CH_3CH_2CH_2CHCH_3 \text{ (with } CH_2CH_3 \text{ branch)}$$

$$CH_3CHCH_2CHCHCH_3 \text{ (with } CH_3 \text{ branches)}$$

Solution:

Locate the longest chain:

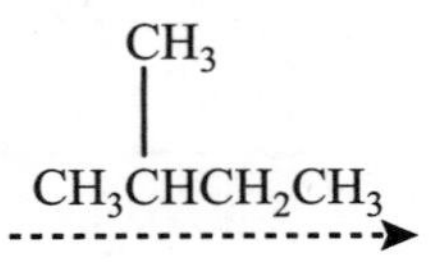

$CH_3CHCH_2CH_3$ with CH_3 branch — 4 carbons, Butane

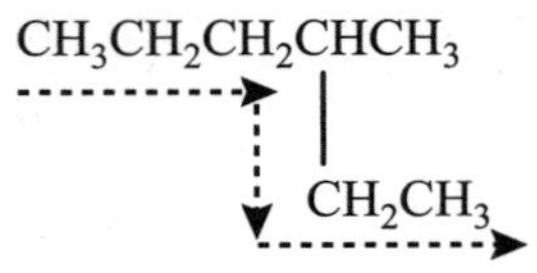

$CH_3CH_2CH_2CHCH_3$ with CH_2CH_3 branch — 6 carbons, Hexane

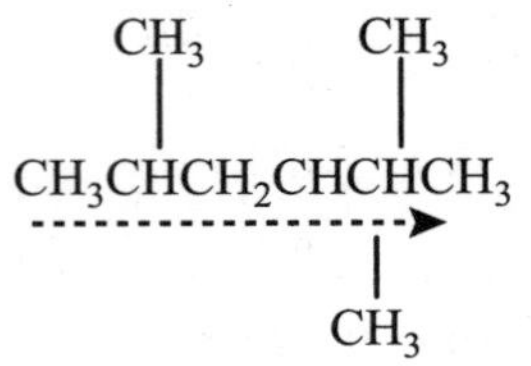

$CH_3CHCH_2CHCHCH_3$ with CH_3 branches — 6 carbons, Hexane

Locate the substituents:

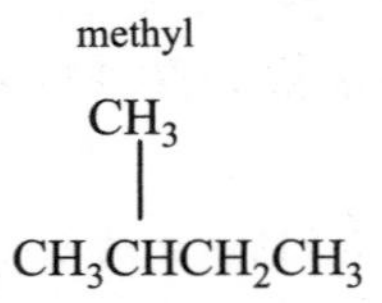

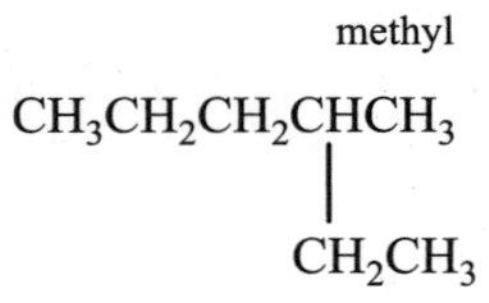

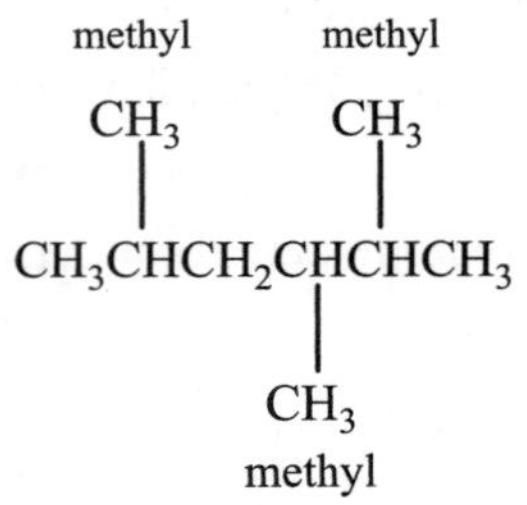

methyl: CH_3 on $CH_3CHCH_2CH_3$

methyl: $CH_3CH_2CH_2CHCH_3$ with CH_2CH_3

methyl, methyl, methyl: CH_3 groups on $CH_3CHCH_2CHCHCH_3$

Number the chain using the lowest numbers:

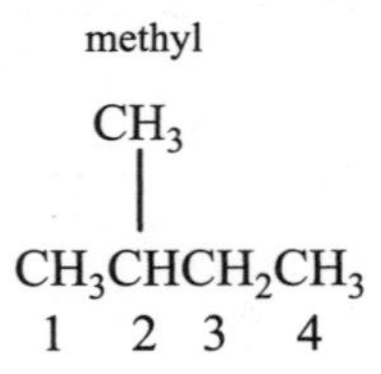

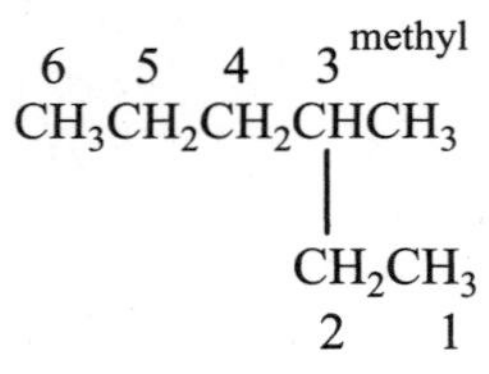

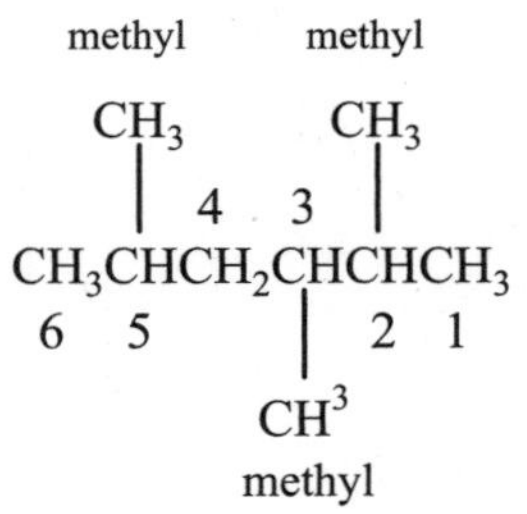

methyl: CH_3 on $CH_3CHCH_2CH_3$ numbered 1 2 3 4

methyl: $CH_3CH_2CH_2CHCH_3$ numbered 6 5 4 3, CH_2CH_3 numbered 2 1

methyl, methyl, methyl: $CH_3CHCH_2CHCHCH_3$ numbered 6 5 4 3 2 1, with CH^3 methyl

Name the molecule:

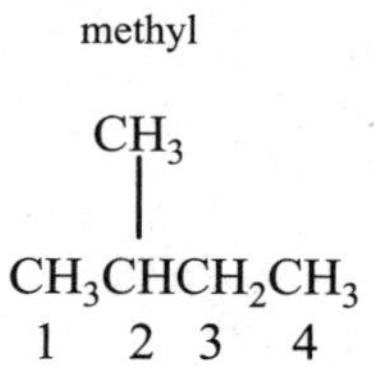

```
methyl

CH3
 |
CH3CHCH2CH3
1   2  3   4

2-methylbutane
```

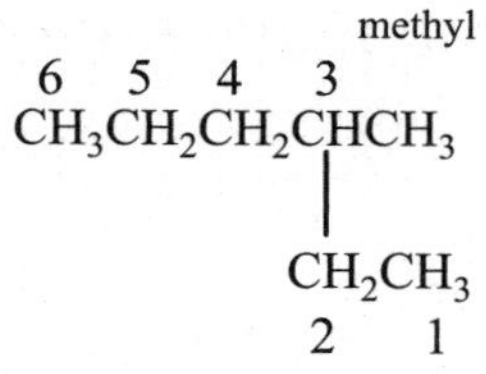

```
                methyl
6   5   4   3
CH3CH2CH2CHCH3
             |
           CH2CH3
           2     1

3-methylhexane
```

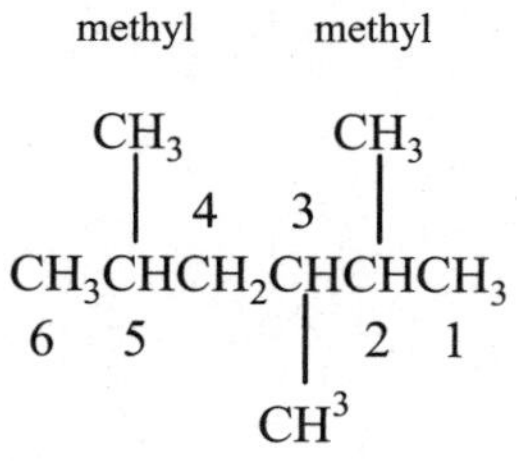

```
methyl      methyl

CH3          CH3
 |   4    3   |
CH3CHCH2CHCHCH3
6   5       |  2  1
           CH3

2,3,5-trimethylhexane
```

Other examples:

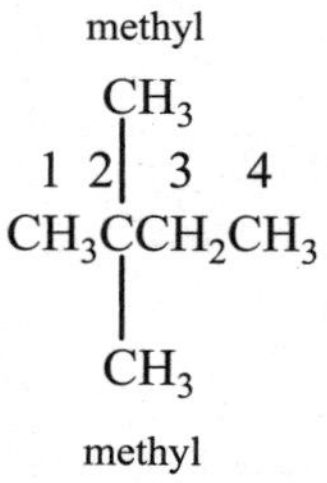

```
  methyl
   CH3
1 2|  3  4
CH3CCH2CH3
   |
   CH3
  methyl

2,2-dimethylbutane

(can use numbers twice)
```

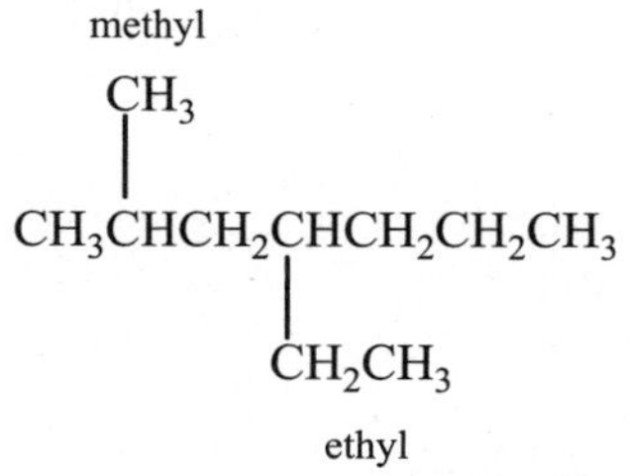

```
methyl
CH3
 |
CH3CHCH2CHCH2CH2CH3
          |
        CH2CH3
         ethyl

4-ethyl-2-methylheptane

(must list groups alphabetically)
```

Practice

Give the IUPAC systematic name for the following molecules.

```
1.    CH2CH2CHCH3
      |       |
  CH3CH2     CH3
```

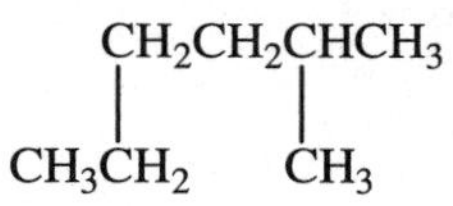

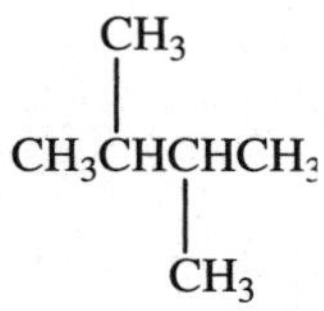

```
      CH3
      |
2. CH3CHCHCH3
         |
        CH3
```

```
          CH3
          |
3. CH3CH2CHCHCH2CH2CH3
            |
          CH2CH3
```

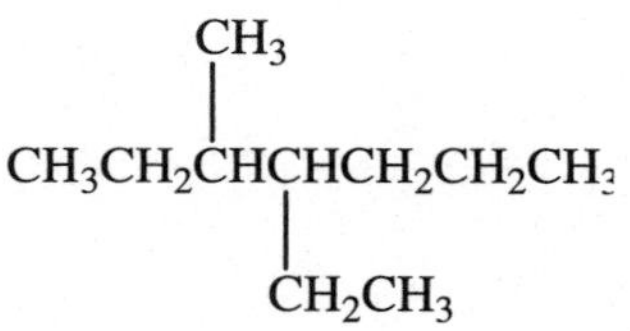

Structural Isomerism

Isomers are defined as different compounds with the same molecular formula. 2-Methylbutane has two other isomers with the molecular formula C_5H_{12}. Each compound must have a unique systematic name. Because molecules can be drawn many different ways, a name can confirm whether the molecule with the same molecular formula is an isomer or the same molecule drawn differently.

$$\begin{array}{ccc} \overset{\displaystyle CH_3}{\overset{|}{CH_3CH_2CHCH_3}} & CH_3CH_2CH_2CH_2CH_3 & \underset{\displaystyle CH_3}{\underset{|}{\overset{\displaystyle CH_3}{\overset{|}{CH_3CCH_3}}}} \end{array}$$

2-methylbutane pentane 2,2-dimethylpropane

Some of the ways to draw 2-methylbutane (i.e., the same molecule drawn differently) are as follows:

$$\overset{\displaystyle CH_3}{\overset{|}{CH_3CH_2CHCH_3}} \qquad \underset{\displaystyle CH_3}{\underset{|}{\overset{\displaystyle CH_3}{\overset{|}{CH_2CHCH_3}}}} \qquad \underset{\displaystyle CH_3}{\underset{|}{\overset{\displaystyle H_3C\ \ CH_3}{\overset{|\quad |}{CHCH_2}}}}$$

Practice

4. Draw the five alkane structural isomers for the molecular formula C_6H_{14}.
5. Name the isomers drawn in problem 4.

Alkenes and Alkynes

The nomenclature of *alkenes* and *alkynes* follows the same rules as alkanes, except the double or triple bond must be numbered. The multiple bond is numbered on the first number to which it is assigned. Also, because double bonds have a rigid configuration, they can exhibit a *cis* or *trans* isomerism. A *cis* structure is one with substituents on the same side of the double bond, and the *trans* is one with the substituents are on opposite sides of the double bond.

Example:

Name the following molecules:

$$\begin{array}{ccc} H_3C \qquad CH_3 & H_3C \qquad H & \\ C{=}C & C{=}C & CH_3CH_2CH_2C{\equiv}CH \\ H \qquad\quad H & H \qquad CH_3 & \end{array}$$

Solution:

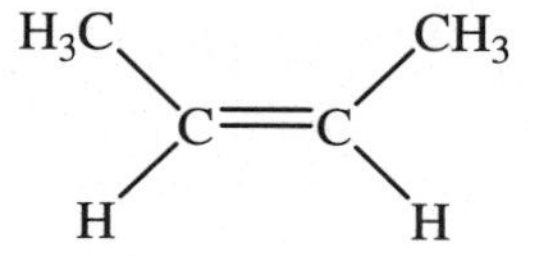

H_3C / CH_3 — C=C — H / H   *cis*-2-butene

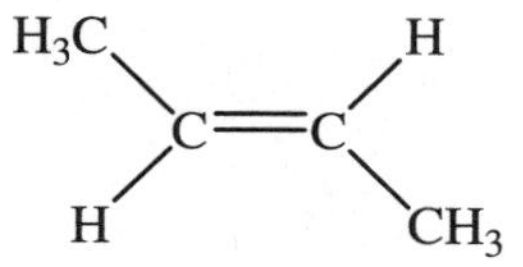

H_3C / H — C=C — H / CH_3 *trans*-2-butene

$CH_3CH_2CH_2C{\equiv}CH$ 1-pentyne

Practice

Give the IUPAC systematic name for the following molecules.

6. $CH_3CH_2CH_2C{\equiv}CCH_2CH_3$

7. H_3C / CH_2CH_3 — C=C — H / H

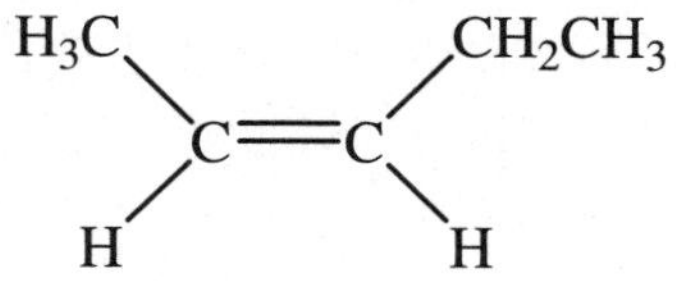

8. H_3CH_2C / H — C=C — H / CH_2CH_3

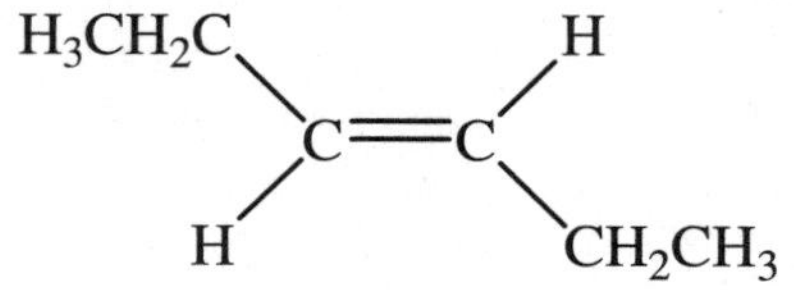

Draw the following molecules:

9. *trans*-3-decene
10. 2-pentyne
11. *cis*-2-octene

Chirality

Take a look at your hands. You can see that they are mirror images of each other. Now, try to orient your right hand so that it perfectly matches your left. You can't, can you? That is because your hands are *chiral*—objects with non-superimposable mirror images. Certain molecules have this property, called chirality, as well. When two molecules have identical connectivity, but cannot be superimposed, they are called *stereoisomers*. In organic chemistry, this occurs most often when a carbon center (*stereocenter*) is bonded to four different substituents. These could be four different atoms, four different carbon chains, or some combination. The molecule bromochloriodomethane offers one of the simplest examples. Look at both of these structures closely. Is there a way to twist them around so they become superimposable?

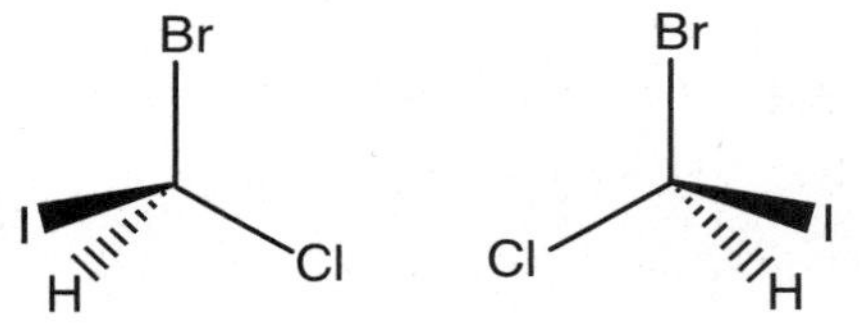

Br, I, H, Cl   Br, Cl, I, H

Chirality is extremely important in the fields of chemistry and biology. All proteins are chiral as well as the amino acids from which they are made. A pharmaceutical chemist developing a new drug must synthesize molecules with the proper chirality for the compound to have the proper effect. The pain reliever ibuprofen has one stereocenter. Despite being nearly identical, only one the isomers below is effective at relieving pain.

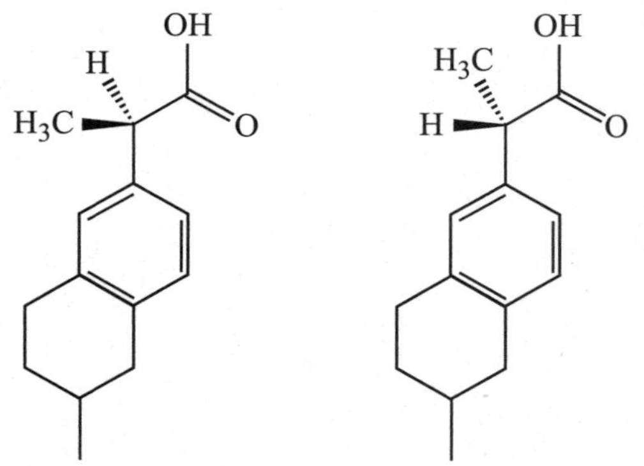

OH, H, H_3C, O   OH, H_3C, H, O

LESSON

21 Organic Chemistry II: Functional Groups

There is more to organic chemistry than carbon and hydrogen. Functional groups such as alcohols, ethers, and carboxylic acids provide a huge amount of diversity to organic molecules. These functional groups are incorporated into organic molecules to produce the industrial chemicals, polymers, and pharmaceuticals we encounter every day.

Alcohols

An *alcohol* is any organic molecule with a hydroxyl group (-OH) appended to it. They have the general formula R-OH. Alcohols are named by taking the name of the corresponding alkane, removing the final -e, and adding -ol to the end.

Table 21.1 Names of alcohols based on the names of their corresponding alkanes

ALKANE FORMULA	ALKANE NAME	ALCOHOL FORMULA	ALCOHOL NAME
CH_4	Methane	CH_3OH	Methanol
CH_3CH_3	Ethane	CH_3CH_2OH	Ethanol
$CH_3CH_2CH_3$	Propane	$CH_3CH_2CH_2OH$	Propanol
$CH_3CH_2CH_2CH_3$	Butane	$CH_3CH_2CH_2CH_2OH$	Butanol

Practice

Identify the following alcohols:

1. 

2. 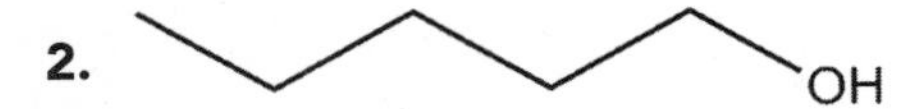

3. 

4. CH_3OH

Ethers

An *ether* is any organic molecule with an oxygen atom linking two alkane chains. They have the formula R-O-R' where R and R' are can either be the same or different alkyl groups. To name them, the shorter alkyl is listed, followed by the longer one, followed by the word *ether*. If the two alkyl groups are identical, the name is only listed once and the prefix *di-* is added. Several examples are given below.

Table 21.2 Names of several types of ethers

CHEMICAL FORMULA	SHORTER ALKYL CHAIN	LONGER ALKYL CHAIN	NAME OF COMPOUND
CH_3OCH_3	Methyl	Methyl	Dimethylether
$CH_3OCH_2CH_3$	Methyl	Ethyl	Methylethylether
$CH_3CH_2OCH_2CH_3$	Ethyl	Ethyl	Diethylether
$CH_3CH_2CH_2OCH_2CH_3$	Ethyl	Propyl	Ethylpropylether

Practice

Identify the following ethers:

5. O

6. O

7. O

8. O

Aldehydes

Aldehydes are organic molecules that possess a terminal carbonyl (C=O) group. They have the formula:

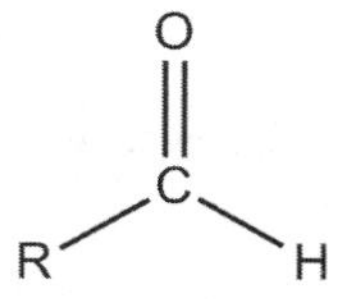

$$R-\overset{\overset{\Large O}{\|}}{C}-H$$

Like for alcohols and ethers, R is any alkyl chain. The naming rules for aldehydes are straightforward: count the number of carbons in the molecule (including the carbonyl carbon), add the appropriate prefix (propyl, butyl, etc), and add the word *aldehyde* at the end. However, the two simplest aldehydes (where R = H and R = CH_3) are exceptions to this naming rule. Instead of using the prefixes methyl- and ethyl -, the prefixes form- and acet- are used instead. These new root words come from an older naming scheme but have stuck around due to common usage.

Practice

Name the following aldehydes:

9. $H-\overset{\overset{\Large O}{\|}}{C}-H$

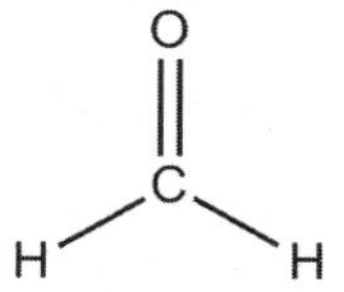

10. $CH_3CH_2-\overset{\overset{\Large O}{\|}}{C}-H$

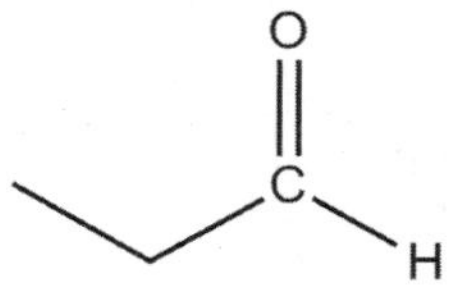

11. $H_3C-\overset{\overset{\Large O}{\|}}{C}-H$

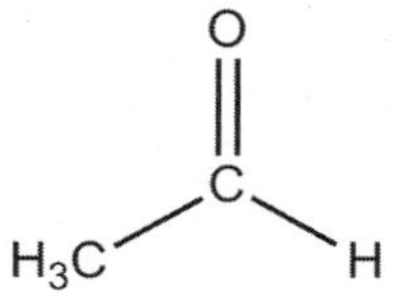

12. $CH_3CH_2CH_2-\overset{\overset{\Large O}{\|}}{C}-H$

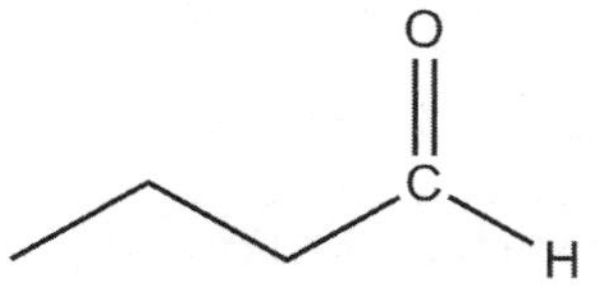

Ketone

A *ketone* is similar to an aldehyde, except that the terminal hydrogen has been replaced with an alkyl group. Ketones have the formula:

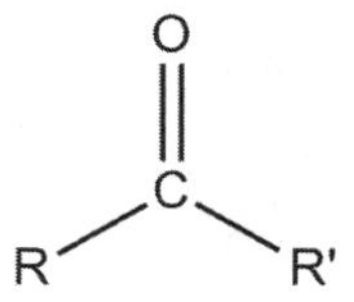

$$R-\overset{\overset{\Large O}{\|}}{C}-R'$$

R and R' can be either the same or different alkyl groups. They are named in a similar manner to ethers: the shorter alkyl group is named, followed by the longer alky group, followed by the suffix –ketone.

Practice

Identify the following ketones:

13.

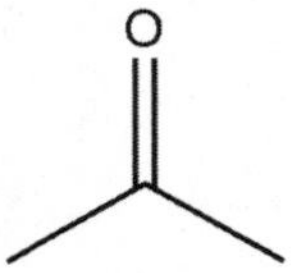

14.

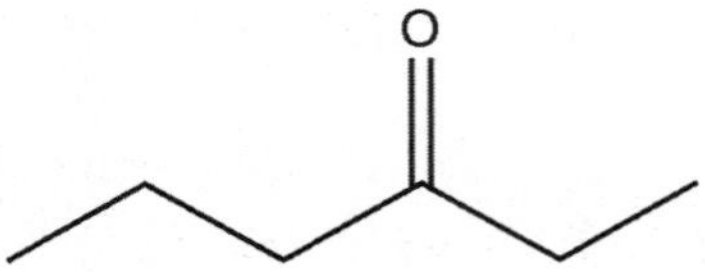

15.

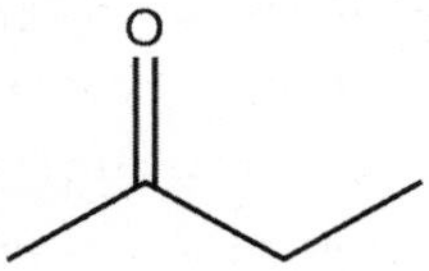

16.

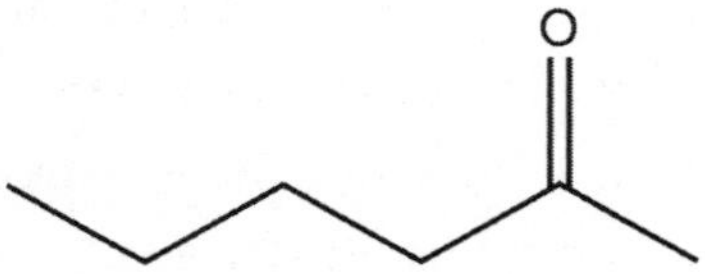

Carboxylic Acids

Carboxylic acids are organic molecules with the formula:

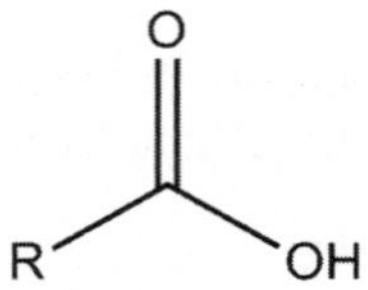

Carboxylic acids are very important in biology. This functional group is what makes an amino acid an acid and without it, amino acids could not link together to form proteins.

These compounds are named using the same prefixes used for alcohols with the suffix *–oic acid* appended to the end. However, as was the case for aldehydes, the two simplest carboxylic acids, (where R = H and R = CH_3) are often known by the more common names, formic acid and acetic acid, respectively. Also, like aldehydes, the prefix specifies the total number of carbons in the molecules (the alkane chain plus the carbonyl carbon).

Practice

Identify the following carboxylic acids:

17.

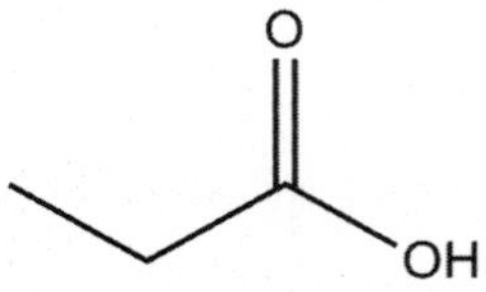

18.

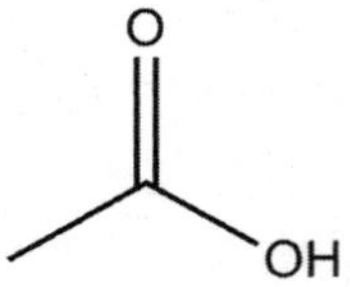

19.

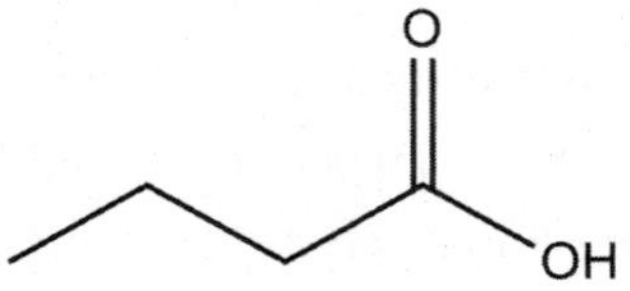

20.

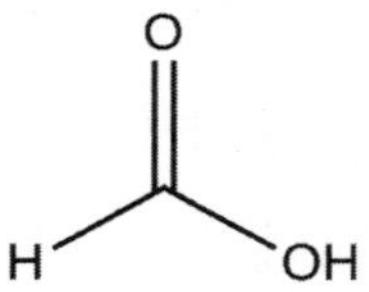

Esters

Esters are alkyl derivatives of carboxylic acids. They have the general formula:

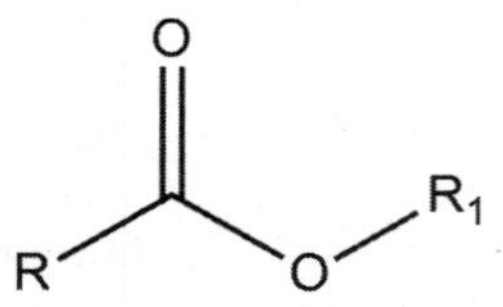

Esters are often quite fragrant and are responsible for the aromas of many fruits and plants. Industrially, esters can be made into long fibers that are used for making clothing (polyester). In general, esters are named by dividing the molecule between its carboxylic acid

(ROOH) and alcohol (R_1OH) parts. Take the alkyl group used to make the alcohol (methyl, ethyl, etc). Next, take the name of the acid and replace the suffix *–ic acid* with *–ate*. Use these two words to form the name of the ester.

Example

Identify the following compound.

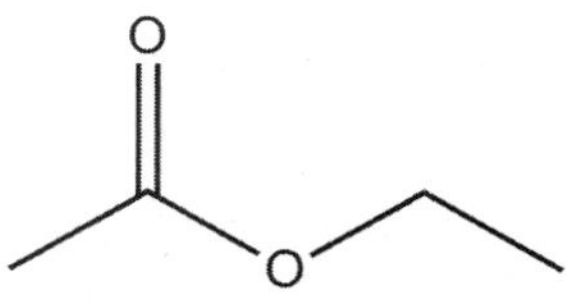

This ester is made from acetic acid and ethanol. So its name is ethyl acetate.

Practice

Identify the following compounds:

21.

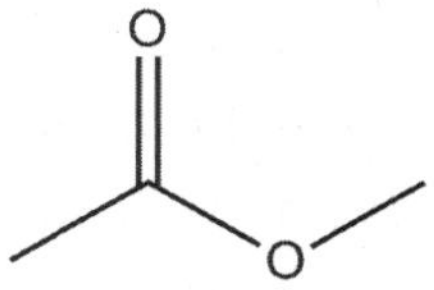

22.

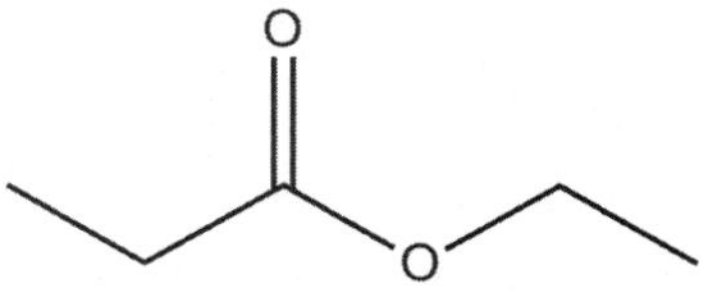

23.

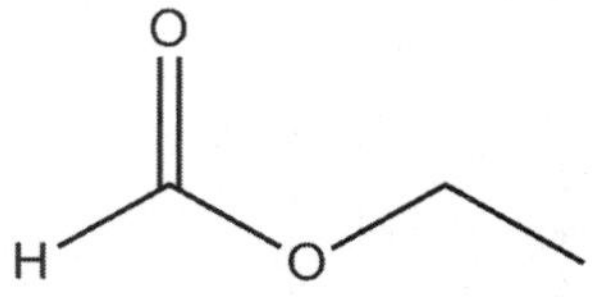

Draw the following compounds:

24. Propyl acetate
25. Ethyl butyrate (naturally occurring in bananas, pineapples, and strawberries)
26. Ethyl pentanoate (naturally occurring in apples)

Amines

Amines are nitrogen-containing organic compounds. They have the general formula:

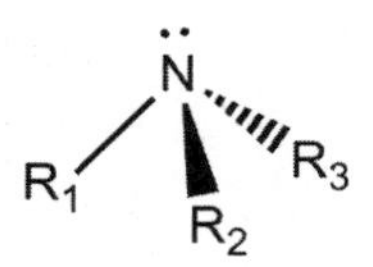

Up to two of the R groups can be replaced with hydrogen atoms (NH_3 is known as ammonia). Amines are classified according to the number of alkyl groups attached to the nitrogen. Amines with only one alkyl group are described as primary (1°), two alkyl groups as secondary (2°), and three alkyl groups as tertiary (3°). Amines are known for their unusually bad odor and are often a major chemical component in the smell of rotting or decaying food. These compounds act as weak bases in many organic syntheses.

The naming scheme for amines is straightforward. The alkyl groups (methyl, ethyl, propyl, etc) are listed, followed by the word amine. If the same alkyl groups appear more than once, the prefix di- or tri- is used.

Example

Identify the amine and state whether it is a primary, secondary, or tertiary amine.

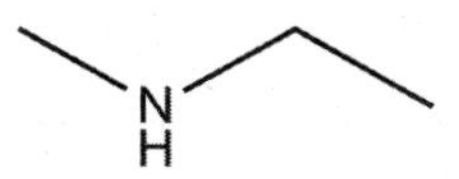

R_1 = methyl
R_2 = ethyl
R_3 = hydrogen
Compound name: Methylethylamine, 2° amine

Practice

Identify the following amines and state whether they are 1°, 2°, or 3° amines:

27. N

28. NH_2

29. N

30. $(CH_3CH_2CH_2)_2NH$

Draw the following organic compounds:

31. Diethylamine
32. Propylamine
33. Methylpropylamine

Other Functional Groups

The functional groups discussed in this chapter are some of the more common groups you will encounter on your journey through chemistry. However, there are many other functional groups that have not been mentioned that are common, both in man-made chemicals as well as in natural products. In fact, nearly all of the drugs used today contain several functional groups within the same molecule. While the naming rules for these molecules are complicated and beyond the scope of this book, it is important to be able to identify these basic groups in more complex organic structures. A table summarizing the functional groups we have discussed as well as a few additional groups is given below.

Table 21.3 Organic Functional Groups

FUNCTIONAL GROUP	GENERAL STRUCTURE	EXAMPLE
Alkane	—C—	CH_4, methane; $CH_3(CH_2)_2CH_3$, butane
Alkene	C=C	$CH_2 = CH_2$, ethane or ethylene
Alkyne	—C≡C—	$HC \equiv CH$, ethyne or acetylene

FUNCTIONAL GROUP	GENERAL STRUCTURE	EXAMPLE
Aromatic	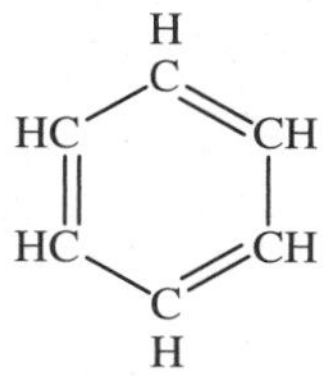 (benzene ring: HC, CH, C–H)	C_6H_6, benzene
Alkyl halide (haloalkane)	RX (X = F, Cl, Br, I)	CH_3CH_2Cl, ethyl chloride
Alcohol	ROH	CH_3CH_2OH, ethyl alcohol
Ether	ROR	$CH_3CH_2OCH_2CH_3$, diethyl ether
Ketone	RCOR or 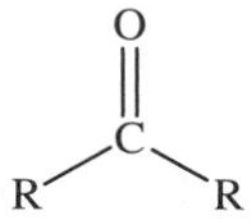 R–C(=O)–R	CH_3COCH_3, acetone
Aldehyde	RCHO or 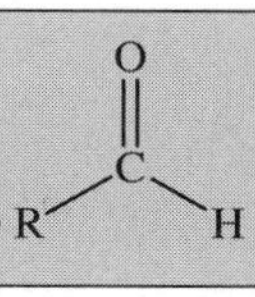 R–C(=O)–H	CH_3CHO, acetaldehyde
Ester	RCO_2R' or R–C(=O)–OR' 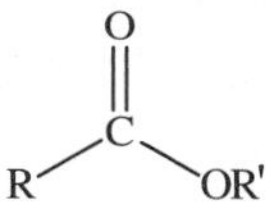	$CH_3CO_2CH_2CH_3$, ethyl acetate
Carboxylic acid	RCO_2H or 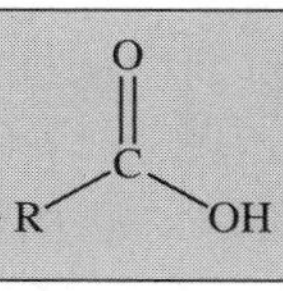 R–C(=O)–OH	CH_3CO_2H, acetic acid
Amine	1°: RNH_2 2°: R_2NH 3°: R_3N 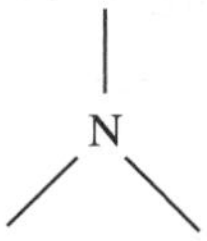	CH_3NH_2, methyl amine $(CH_3)_2NH$, dimethyl amine $(CH_3)_3N$, trimethyl amine
Amide	$RCONH_2$ or 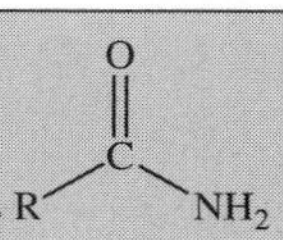 R–C(=O)–NH_2	CH_3CONH_2, acetamide
Nitrile	RC ≡ N	$CH_3C \equiv N$, acetonitrile
Imine	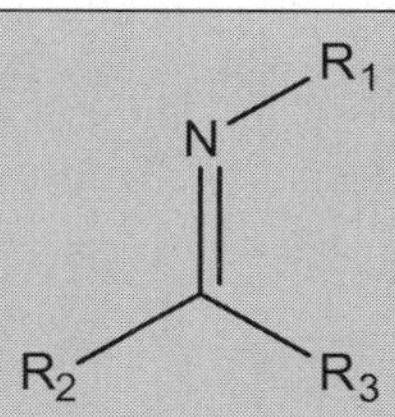 R_2, R_3–C=N–R_1	$(CH_3)_2CNH$, propanoamine
Thioether	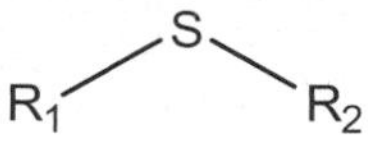 R_1–S–R_2	$(CH_3)_2S$, dimethyl thioether

Example

Identify the functional groups in phenylalanine (an amino acid and a component in aspartame and proteins) and vanillin (vanilla scent).

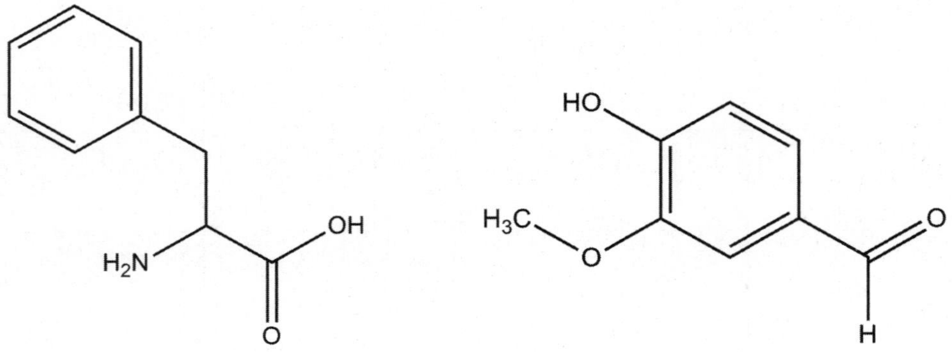

OH
H_2N
O
HO
H_3C
O
O
H

Solution

Aromatic
Carboxylic Acid
OH
H_2N
1° Amine
O
Alcohol
HO
H_3C
O
O
Ether
Aldehyde
H

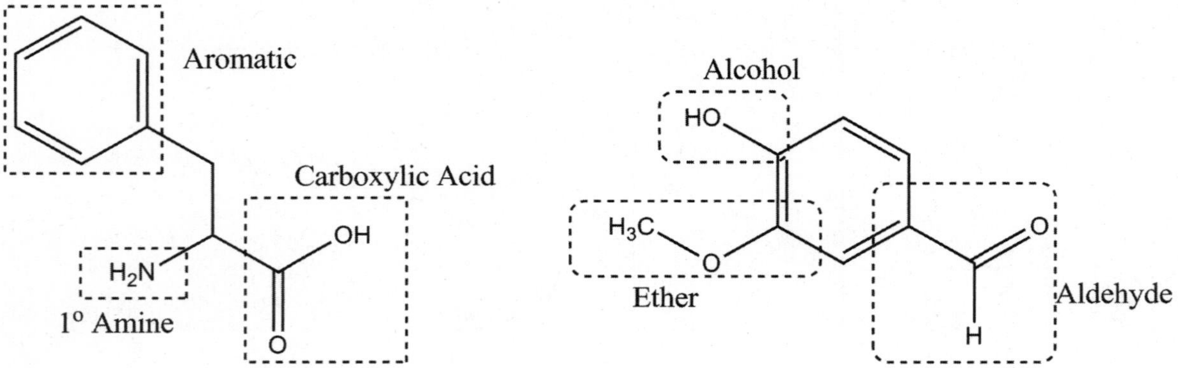

Practice

Identify the functional groups in the following organic compounds:

34. Methyl anthranilate (responsible for the aroma of grapes)

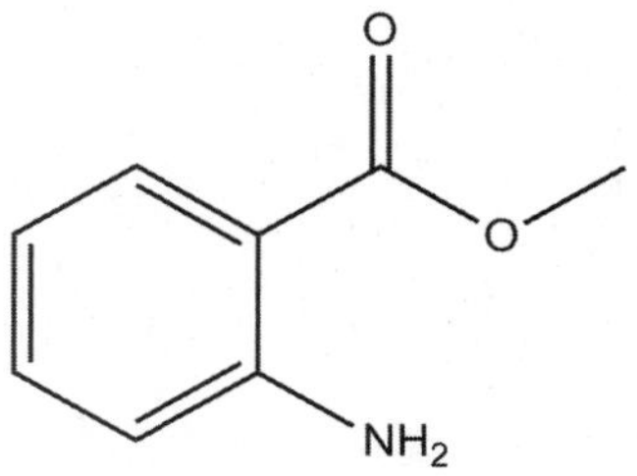

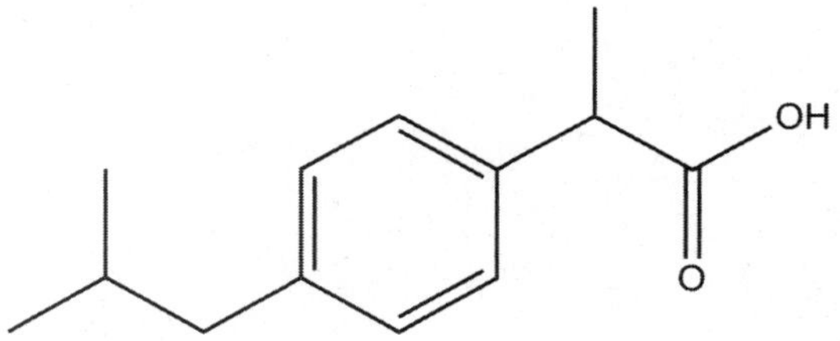

36. Clindamycin (antibiotic)

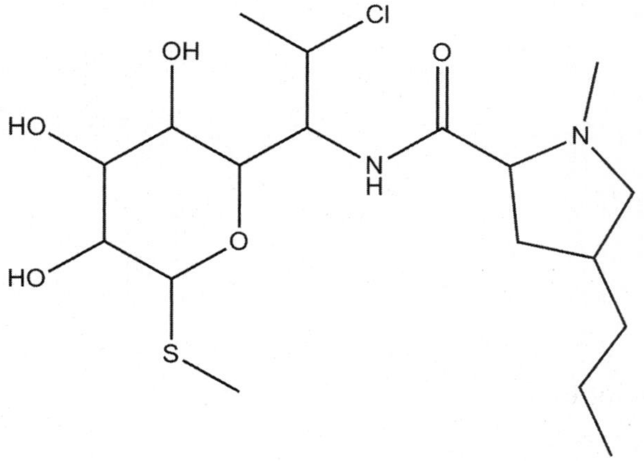

35. Ibuprofen (a common pain reliever)

37. Penicillin (antibiotic)

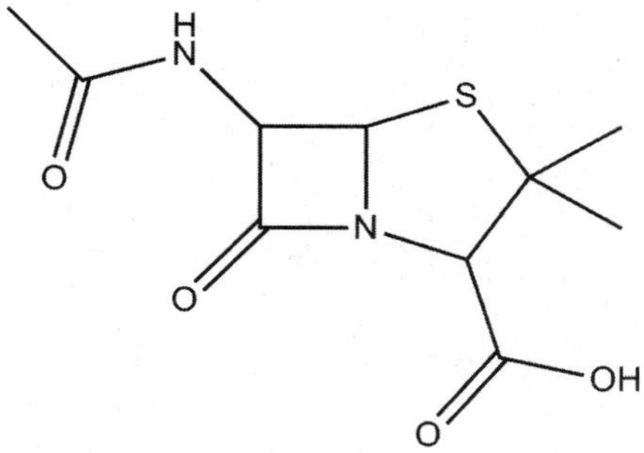

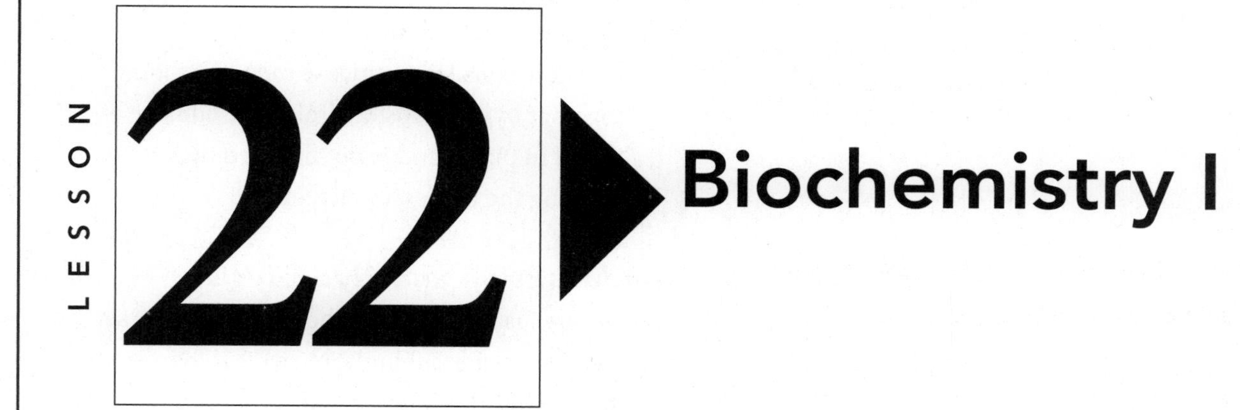

Biochemistry I

Biochemistry is the chemistry of life. Biological systems have developed specialized molecules and chemical reactions to store genetic information and regulate the processes of life. While biomolecules may have a different appearance than the organic molecules studied in previous lessons, they are assembled from the same functional groups you are already familiar with.

Carbohydrates

Carbohydrates (or sugars) serve as the main source of energy for living organisms. They are composed of carbon, hydrogen, and oxygen with the general molecular formula $C_x(H_2O)_y$, a hydrate of carbon. Carbohydrate names have the suffix *–ose* (e.g., such as gluc*ose* or fruct*ose*).

Monosaccharides

Monosaccharides are the simplest carbohydrate structure composed of rings or chains containing five C atoms (*pent*ose, such as ribose, which is a constituent of RNA) or six C atoms (*hex*ose, such as galactose derived from milk sugar lactose).

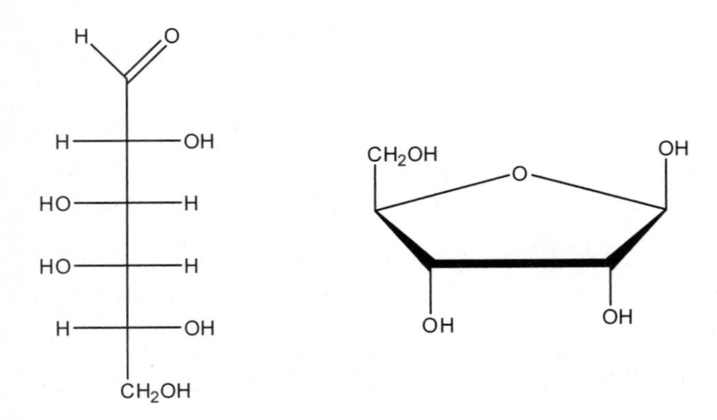

Figure 22.1. A linear chain of galactose and ribose, a five carbon monosaccharide

Disaccharides

Disaccharides are dimeric sugars made of two monosaccharides joined together in a reaction that releases a molecule of water (dehydration). The bond between the two sugar molecules is called *glycosidic linkage* and can have either an axial (α-glycoside) or an equatorial (β-glycoside) orientation with respect to the ring conformation. Examples include the following:

- *Maltose* (two glucose molecules joined together), found in starch
- *Lactose* (one galactose joined to one glucose), found in milk
- *Sucrose* (one fructose joined to one glucose), table sugar

Polysaccharides

Polysaccharides are polymers or a long chain of repeating monosaccharide units. Examples include the following:

- *Starch* is a mixture of two kinds of polymers of α-glucose (linear amylose and amylopectin).
- *Amylose* contains glucose molecules joined together by α-glycosidic linkages, and amylopectin additionally has a branching at C-6. They are storage polysaccharides in plants.
- *Glycogen* consists of glucose molecules linked by a α-glycosidic linkage (C-1 and C-4) and branched (C-6) by α-glycosidc linkage. Glycogen is the storage form of glucose in animals (liver and skeletal muscle).
- *Cellulose* consists of glucose molecules joined together by β-glycosidic linkage. Cellulose is found in plants and is not digested by humans (lacking the necessary enzyme).

Condensation and Hydrolysis

Condensation is the process of bonding together separate monosaccharide subunits into a disaccharide and/or a polysaccharide. It is also called *dehydration synthesis* as one molecule of water is lost in the process. It is carried out by specific enzymes.

Hydrolysis is the reverse process of condensation as a water molecule and specific enzymes break all the glycosidic linkages in disaccharides and polysaccharides into their constituting monosaccharides.

Lipids

Lipids are a diverse group of compounds that are insoluble in water and polar solvents but are soluble in nonpolar solvents. Lipids are stored in the body as a source of energy (twice the energy provided by an equal amount of carbohydrates).

Triglycerides

Triglycerides are lipids formed by the condensation of glycerol (one molecule) with fatty acids (three molecules). They can be saturated (from fatty acid containing only C-C single bonds) or unsaturated (the presence of one or more C=C double bonds). Triglycerides are found in the adipose cells of the body (neutral fat) and are metabolized by the enzyme lipase (an esterase) during hydrolysis, producing fatty acids and glycerol.

Ketone Bodies

Three *ketone bodies* are formed during the breakdown (metabolism) of fats: acetoacetate, β-hydroxybutyrate, and acetone. They are produced to meet the energy requirements of other tissues. Fatty acids, produced by the hydrolysis of triglycerides, are converted to ketone bodies in the liver. They are removed by the kidneys *(ketosuria)*, but if found in excess in the blood *(ketonemia)*, ketone bodies can cause a decrease of the blood pH and *ketoacidosis* may result. The ketone body acetone is exhaled via the lungs (this process is called *ketosis*). Ketosuria and ketonemia are common in diabetes mellitus patients and in cases of prolonged starvation.

Phospholipids

Phospholipids are lipids containing a phosphate group. They are the main constituents of cellular membranes.

Steroids

Steroids are organic compounds characterized by a core structure known as *gonane* (three cyclohexane or six-carbon rings and one cyclopentane or a five-carbon ring fused together). Steroids differ by the functional groups attached to the gonane core. Cholesterol is an example of a steroid and is a precursor to the steroid hormones such as the sex hormones (the androgens and estrogens) and the corticosteroids (hormones of the adrenal cortex).

Figure 22.2 The chemical structure of estrogen

Table 22.1 The Naturally Occurring Amino Acids

AMINO ACID	ABBREVIATION	R GROUP
Alanine	Ala	$-CH_3$
Arginine	Arg	$-CH_2-CH_2-CH_2-NH-C(=NH)-NH_2$
Asparagine	Asn	$-CH_2-C(=O)-NH_2$
Aspartic acid	Asp	$-CH_2-C(=O)-OH$
Cysteine	Cys	$-CH_2-SH$
Glutamic acid	Glu	$-CH_2-CH_2-C(=O)-OH$
Glutamine	Gln	$-CH_2-CH_2-C(=O)-NH_2$
Glycine	Gly	$-H$

Table 22.1 (continued)

AMINO ACID	ABBREVIATION	R GROUP
Histidine	His	$-CH_2-$ (ring with NH, N)
Isoleucine	Ile	$-CH(CH_3)-CH_2-CH_3$
Leucine	Leu	$-CH_2-CH(CH_3)-CH_3$
Lysine	Lys	$-CH_2-CH_2-CH_2-CH_2-NH_2$
Methionine	Met	$-CH_2-CH_2-S-CH_3$
Phenylalanine	Phe	$-CH_2-$ (benzene ring)
Proline	Pro	$O=C(OH)-$ (ring with NH)
Serine	Ser	$-CH_2-OH$
Threonine	Thr	$-CH(OH)-CH_3$
Tryptophan	Trp	$-CH_2-$ (ring with NH)
Tyrosine	Tyr	$-CH_2-$ (benzene ring) $-OH$
Valine	Val	$-CH(CH_3)-CH_3$

Proteins

Every organism contains thousands of different proteins with a variety of functions: structure (collagen or histones), transport (hemoglobin or serum albumin), defense (the antibodies or fibrinogen for blood coagulation), control and regulation (insulin), catalysis (the enzymes), and storage. *Proteins* (also called *polypeptides*) are long chains of amino acids joined together by covalent bonds of the same type (peptide or amide bonds). Twenty naturally occurring amino acids exist, each characterized by an amino group at one end and a carboxylic acid group at the other end. Different proteins are characterized by different amino acids and/or a difference order of amino acids.

The sequence of amino acids in the long chain defines the *primary structure* of a protein. A *secondary structure* is determined when several residues, linked by hydrogen bonds, conform to a given combination (e.g., the α-helix, pleated sheet, and β-turns). *Tertiary structure* refers to the three-dimensional folded conformation of a protein. This is the biologically active conformation. A *quaternary structure* can result when two or more individual proteins assemble into two or more polypeptide chains. *Conjugated proteins* are complexes of proteins with other biomolecules, such as glycoproteins (sugar-proteins).

Enzymes

Enzymes are biological catalysts whose role is to increase the rate of chemical (metabolic) reactions without being consumed in the reaction. They do so by lowering the activation energy of a reaction by binding specifically (i.e., in the active site) to their substrates in a "lock-and-key" or "induced-fit" mechanism. They do not change the nature of the reaction (in fact, any change is associated with a malfunctioning enzyme), the onset of a disease, or its outcome.

Enzyme activity is influenced by:

- **Temperature:** Proteins can be destroyed at high temperatures and their action is slowed at low temperature.
- **pH:** Enzymes are active in a certain range of the pH.
- **Concentration of enzymes and substrates.**
- **Concentration of cofactors and coenzymes (vitamins).**

Protein Denaturation

Protein denaturation occurs when the protein configuration is changed by the destruction of the secondary and tertiary structures (reduced to the primary structure). Common denaturing agents are alcohol, heat, and heavy metal salts.

Practice

1. What three elements are found in carbohydrates?
2. What is the primary function of food carbohydrates in the body?
3. What are steroids classified as?
4. A high level of ketone bodies in urine indicates a marked increase in the metabolism of what?
5. What molecules are produced when sucrose is broken down by the enzyme sucrase?
6. What are the bonds between amino acids in a polypeptide called?
7. Which polysaccharide is a branched polymer of α-glucose found in the liver and muscle cells?
8. What is the site on an enzyme molecule that does the catalytic work?

LESSON

23 Biochemistry II: Nucleic Acids

So far, we have discussed the major compounds—proteins, lipids, and carbohydrates—that make up the components of a cell. But how do living things store the vast amount of information required to synthesize these compounds? This lesson describes the chemistry of DNA and RNA and how this information is used to store, transmit, and record what is essentially the code of life.

Nucleic Acids

The ability of an organism to regulate the numerous concurrent processes taking place requires a vast amount of information. This information must not only be easy to access, it must be readily able to be copied during cell division. All forms of life use chains of deoxyribonucleic acid (DNA) to store genetic material.

This genetic code is written with four letters: A, T, G, and C. These letters represent the four *nucleosides* (adenine, thymine, cytosine, and guanine) that bond together to form chains known as *nucleic acids*. Nucleosides are composed of a sugar ring (2-deoxyribose), bonded to a specific amine base at the 1' position. Nucleosides can be linked together through the formation of a phosphoric ester. Chains of *nucleotides* form single-stranded DNA.

Figure 23.1 A dinucleotide containing thymine and guanine

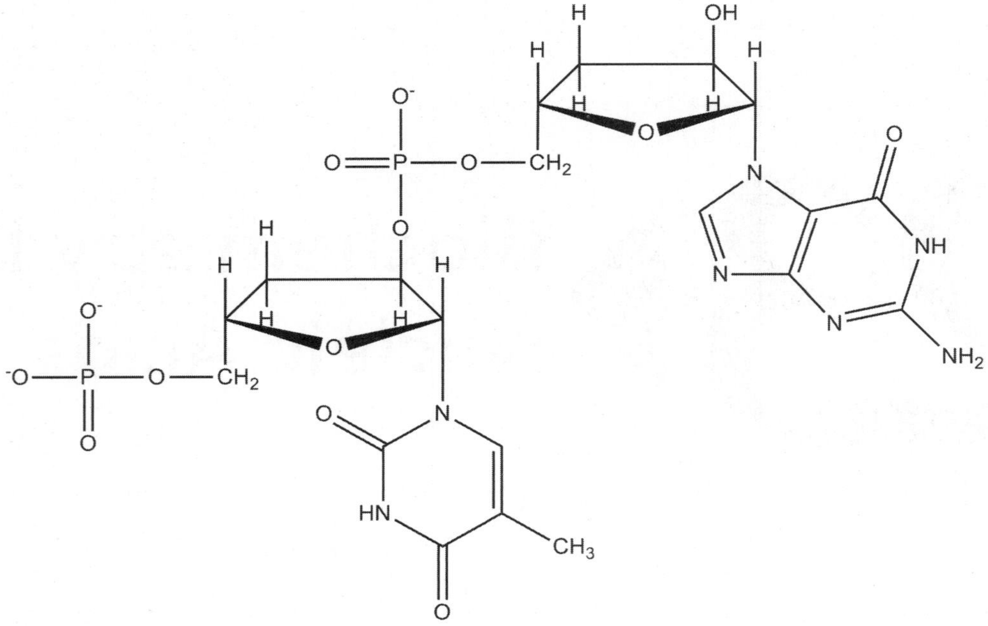

Table 23.1 The four nucleotide bases found in DNA

NAME OF BASE	ABBREVIATION	STRUCTURE
Adenine	A	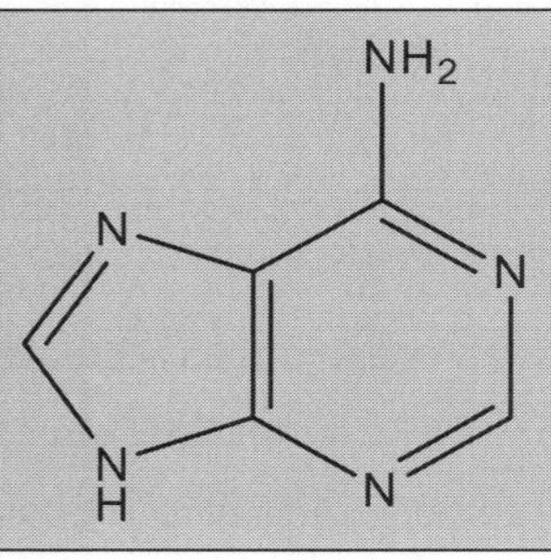
Cytosine	C	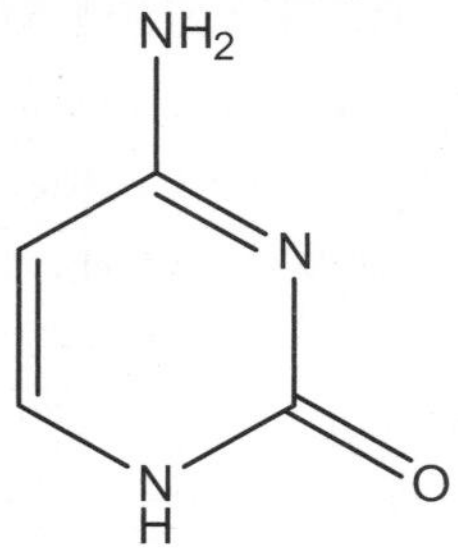

NAME OF BASE	ABBREVIATION	STRUCTURE
Guanine	G	
Thymine	T	

Base Pairing

In 1953, American and British scientists proposed a novel idea for the structure of DNA and how this structure could contribute to the transfer of genetic information. In 1962, these scientists won the Nobel Prize for the discovery of the *double helical structure* of DNA. Due to precisely placed hydrogen bonds, adenine is only able to form a *base pair* with thymine. In addition, cytosine is only able to form a base pair with guanine. Single DNA strands pair with a complementary strands (adenines are matched with thymines and guanines are matched with cytosines). In DNA replication, the two stands separate and individual nucleotides bond to each strand, creating a copy of the original and allowing genetic material to be passed on.

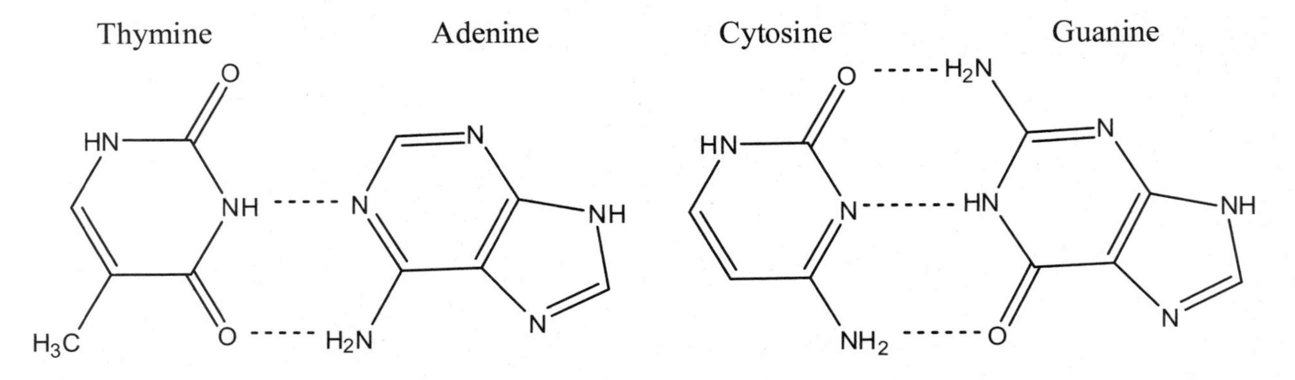

Fig. 23.2 Base pairing between adenine and thymine, cytosine and guanine

Practice

1. What is the complementary sequence to the DNA strand ATGCTCAACGAGCGATAA?
2. What is the complementary sequence to the DNA strand ATGGAACTTCATTTTTAG?

Transcription and Translation

DNA contains the code to synthesize all the proteins expressed in an organism. But how is the genetic information contained in DNA used to create proteins? In the first step, *transcription*, the information in DNA is copied to a complementary ribonucleic acid (RNA) strand. In RNA, the base thymine is replaced with another amine base, uracil (U). This RNA strand is then used as the template for protein synthesis in a process known as *translation*. In translation, groups of three RNA nucleotides (codons) are used to identify one of twenty amino acids. In addition, there are codons to start and stop translation. These amino acids combine to form polypeptides (proteins). Mutations arise when individual DNA nucleotides are altered, leading to changes in the transcribed proteins.

Table 23.2 RNA Codons and their corresponding amino acids

FIRST POSITION	SECOND POSITION								THIRD POSITION
	U		C		A		G		
U	UUU	Phe	UCU	Ser	UAU	Tyr	UGU	Cys	U
	UUC	Phe	UCC	Ser	UAC	Tyr	UCG	Cys	C
	UUA	Leu	UCA	Ser	UAA	Stop	UGA	Stop	A
	UUG	Leu	UCG	Ser	UAG	Stop	UGG	Trp	G
C	CUU	Leu	CCU	Pro	CAU	His	CGU	Arg	U
	CUC	Leu	CCC	Pro	CAC	His	CGC	Arg	C
	CUA	Leu	CCA	Pro	CAA	Gln	CGA	Arg	A
	CUG	Leu	CCG	Pro	CAG	Gln	CGG	Arg	G
A	AUU	Ile	ACU	Thr	AAU	Asn	AGU	Ser	U
	AUC	Thr	AAC	Asn	AAC	Asn	AGC	Ser	C
	AUA	Ile	ACA	Thr	AAA	Lys	AGA	Arg	A
	AUG*	Met	ACG	Thr	AAG	Lys	AGG	Arg	G
G	GUU	Val	GCU	Ala	GAU	Asp	GGU	Gly	U
	GUC	Val	GCC	Ala	GAC	Asp	GGC	Gly	C
	GUA	Val	GCA	Ala	GAA	Glu	GGA	Gly	A
	GUG	Val	GCG	Ala	GAG	Glu	GGG	Gly	G

*AUG is also the 'start' codon that initiates translation.

Practice

3. Translate the following DNA sequence into the complementary RNA strand: AGTAACCAG.
4. Translate the following RNA sequence into a protein sequence: AUGCUCAACGAGCGAUAA.
5. Translate the following RNA sequence into a protein sequence: AUGGAACUUCAUUUUUAG.

LESSON

24 Nuclear Processes

A *nuclear reaction* is a reaction at the atomic level where energy and/or mass are released or absorbed in a process. Nuclear medicine is one of the fastest-growing branches of medicine, and has proven to be beneficial in the detection and treatment of many diseases, including cancer.

Nuclear Reactions

Nuclear chemistry describes reactions involving changes in atomic nuclei. In Lesson 2, elements were defined as matter that cannot be broken down by simple means. Some isotopes are radioactive and are broken down by nuclear processes. *Radioactivity* is the process by which unstable nuclei break down spontaneously, emitting particles and/or *electromagnetic radiation* (i.e., energy), also called *nuclear radiation*. Heavy elements (from atomic number 83) are naturally radioactive, and many more (the transuranium elements, atomic numbers 93 to 116) have been generated in laboratories.

Spontaneous Nuclear Processes

Spontaneous nuclear processes include the following:

- **Alpha emission:** An alpha particle (symbol: $^{4}_{2}He$ or $^{4}_{2}\alpha$) corresponds to the nucleus of a helium atom (having two protons and two neutrons) that is spontaneously emitted by a nuclear breakdown or decay. The α-particles are of low energy and therefore low penetrating (a lab coat is sufficient to block their penetration), but dangerous if inhaled or ingested.

$$^{234}_{92}U \rightarrow ^{230}_{90}Th + ^{4}_{2}He$$

- **Beta emission:** A beta particle (symbol: $^{0}_{-1}\beta$ or $^{0}_{-1}e$) is an electron released with high speed by a radioactive nucleus, in which neutrons (in excess) are converted into protons and electrons (i.e., β-particles). The β-particles are medium-penetrating radiation requiring dense material and several layers of clothing to block their penetration. They are dangerous if inhaled or ingested.

$$^{14}_{6}C \rightarrow ^{14}_{7}N + ^{0}_{-1}e$$

- **Gamma emission:** Gamma rays (symbol: γ) are massless and chargeless forms of radiation (pure energy). They are the most penetrating form of radiation, similar to X-rays, and can only be stopped by barriers of heavy materials such as concrete, lead, and so on. They are extremely dangerous and can cause damage to the human body.

Nonspontaneous Nuclear Processes

In addition, neutrons ($^{1}_{0}n$) can be used to bombard a nucleus, and neutrons can be products of nuclear processes.

Nuclear transmutation is another type of radioactivity occurring when nuclei are bombarded by other particles (protons or neutrons) or nuclei. By this process, lighter elements can be enriched and thus converted to heavier ones or vice versa. Lord Rutherford observed the first transformation in the early twentieth century:

$$^{14}_{7}N + ^{4}_{2}He \rightarrow ^{17}_{8}O + ^{1}_{1}H$$

Notice in the example that the atomic number (protons) adds to nine on each side of the equation and the mass numbers (protons + neutrons) add to 18 on both sides of the equation. During a nuclear reaction, the following occurs:

- **conservation of mass number** (i.e., the same number of protons in the products and reactants)
- **conservation of atomic number** (i.e., the same number of protons and neutrons in the products and reactants)

The two principle types of nuclear reactions are fission and fusion. *Nuclear fusion* is the process in which small nuclei are combined (i.e., fused) into larger (more stable) ones with the release of a large amount of energy. Fusion reactions take place at very high temperatures and pressures (thermonuclear reactions) as it occurs in the sun. Examples of fusion include the following:

$$^{2}_{1}H + ^{1}_{1}H \rightarrow ^{3}_{2}He + \gamma$$

$$^{13}_{6}C + ^{4}_{2}He \rightarrow ^{16}_{8}O + ^{1}_{0}n$$

All elements are products of solar fusion reactions.

Nuclear fission is the process in which a heavier nucleus (usually less stable) splits into smaller nuclei and neutrons. The process releases a large amount of energy and neutrons that can set up a *chain reaction* (or self-sustaining nuclear fission reactions) with more and more releases of energy (highly exothermic reactions) and neutrons. Examples of fission include the following:

$$^{236}_{92}U \rightarrow ^{103}_{42}Mo + ^{131}_{50}Sn + 2^{1}_{0}n$$

A *radioactive isotope (radioisotope)* is an unstable isotope of an element that decays into a more stable isotope of the same element. They are of great use in medicine as tracers (to help monitor particular atoms in chemical and biological reactions) for the purpose of diagnosis (such as imaging) and treatment. Iodine (-131 and -123) and Technetium-99 are used for their short half-lives.

Practice

1. What is least penetrating radiation given off a radioactive substance?
2. If $^{238}_{92}U$, an isotope of uranium gives off a beta-particle and gamma rays, what is the resulting isotope?
3. What is the missing product of $^{42}_{17}Cl \rightarrow ^{42}_{18}Ar + ?$
4. What is the missing product of $^{60}_{24}Cr \rightarrow ^{56}_{22}Ti + ?$
5. What is the missing product of $^{192}_{78}Pt \rightarrow ^{188}_{76}Os + ?$
6. What is the missing product of $^{241}_{96}Cm + ? \rightarrow ^{241}_{95}Am$?
7. What are the two types of nuclear reactions?

Half-Life

A radioactive isotope's *half-life* (symbol: $t_{\frac{1}{2}}$) is the time required for the concentration of the nuclei in a given sample to decrease to half its initial concentration. The half-life is specific to a radioactive element and varies widely (from three hours for Sr-87 to millions of years for U-238, for example). The mathematical expression for half-life is as follows (k is the rate constant):

$$t_{\frac{1}{2}} = \frac{0.693}{k}$$

Example:
Technetium-99 ($^{99}_{43}Tc$) is a common radioisotope used in nuclear medicine. The rate constant for Tc-99 is 1.16×10^{-1}/h. What is the half-life of Tc-99?

Solution:

$$t_{\frac{1}{2}} = \frac{0.693}{k} = \frac{0.693}{1.16 \times 10^{-1}/h} = 5.97 \text{ hours}$$

Example:
The half-life of a given element is 70 years. How long will it take 5.0 g of this element to be reduced to 1.25 g?

Solution:
The easiest way to solve this problem is to recognize that two half-lives will be needed ($5.0 \rightarrow 2.5 \rightarrow 1.25$) for this decay: 70 years $\times$ 2 = 140 years.

Practice

What is the half-life for the following isotopes?

8. $^{90}_{38}Sr$, with a rate constant of 0.0241 h^{-1}.
9. $^{131}_{53}I$, with a rate constant of 0.0856 $days^{-1}$.
10. The half-life of Phosphorus-32 is 14.3 days. How long will it take 16.0 g of this isotope to be reduced to 2.0 g?

POSTTEST

If you have completed all 20 lessons in this book, then you are ready to take the posttest to measure your progress. The posttest has 30 multiple-choice questions covering the topics you studied in this book. Although the format of the posttest is similar to that of the pretest, the questions are different.

Take as much time as you need to complete the posttest. When you are finished, check your answers with the answer key at the end of the posttest section. Along with each answer is the lesson that covers the chemistry skills needed for that question. Once you know your score on the posttest, compare the results with the pretest. If you scored better on the posttest than you did on the pretest, you should look at the questions you missed, if any. Do you know why you missed the question, or do you need to go back to the lesson and review the concept?

If your score on the posttest does not show much improvement, take a second look at the questions you missed. Did you miss a question because of an error you made? If you can figure out why your answer was incorrect, then you understand the concept and just need to concentrate more on accuracy when taking a test. If you missed a question because you did not know how to work the problem, go back to the lesson and spend more time working on that type of problem. Take the time to understand basic chemistry thoroughly. You need a solid foundation in basic chemistry if you plan to use this information or progress to a higher level. Whatever your score on this posttest, keep this book for review and future reference.

Answer Sheet

1.	ⓐ	ⓑ	ⓒ	ⓓ	11.	ⓐ	ⓑ	ⓒ	ⓓ	21.	ⓐ	ⓑ	ⓒ	ⓓ
2.	ⓐ	ⓑ	ⓒ	ⓓ	12.	ⓐ	ⓑ	ⓒ	ⓓ	22.	ⓐ	ⓑ	ⓒ	ⓓ
3.	ⓐ	ⓑ	ⓒ	ⓓ	13.	ⓐ	ⓑ	ⓒ	ⓓ	23.	ⓐ	ⓑ	ⓒ	ⓓ
4.	ⓐ	ⓑ	ⓒ	ⓓ	14.	ⓐ	ⓑ	ⓒ	ⓓ	24.	ⓐ	ⓑ	ⓒ	ⓓ
5.	ⓐ	ⓑ	ⓒ	ⓓ	15.	ⓐ	ⓑ	ⓒ	ⓓ	25.	ⓐ	ⓑ	ⓒ	ⓓ
6.	ⓐ	ⓑ	ⓒ	ⓓ	16.	ⓐ	ⓑ	ⓒ	ⓓ	26.	ⓐ	ⓑ	ⓒ	ⓓ
7.	ⓐ	ⓑ	ⓒ	ⓓ	17.	ⓐ	ⓑ	ⓒ	ⓓ	27.	ⓐ	ⓑ	ⓒ	ⓓ
8.	ⓐ	ⓑ	ⓒ	ⓓ	18.	ⓐ	ⓑ	ⓒ	ⓓ	28.	ⓐ	ⓑ	ⓒ	ⓓ
9.	ⓐ	ⓑ	ⓒ	ⓓ	19.	ⓐ	ⓑ	ⓒ	ⓓ	29.	ⓐ	ⓑ	ⓒ	ⓓ
10.	ⓐ	ⓑ	ⓒ	ⓓ	20.	ⓐ	ⓑ	ⓒ	ⓓ	30.	ⓐ	ⓑ	ⓒ	ⓓ

1. In a dilute solution of sodium chloride in water, the sodium chloride is the
   a. solvent.
   b. solute.
   c. precipitate.
   d. reactant.

2. A sample of nitrogen at 20° C in a volume of 875 mL has a pressure of 730 mm Hg. What will be its pressure at 20° C if the volume is changed to 955 mL?
   a. 750 mm Hg
   b. 658 mm Hg
   c. 797 mm Hg
   d. 669 mm Hg

3. A mixture consisting of 8.0 g of oxygen and 14 g of nitrogen is prepared in a container such that the total pressure is 750 mm Hg. The partial pressure of oxygen in the mixture is
   a. 125 mm Hg.
   b. 500 mm Hg.
   c. 135 mm Hg.
   d. 250 mm Hg.

4. To prepare 100 ml of 0.20 M NaCl solution from stock solution of 1.00 M NaCl, you should mix
   a. 20 mL of stock solution with 80 mL of water.
   b. 40 mL of stock solution with 60 mL of water.
   c. 20 mL of stock solution with 100 mL of water.
   d. 25 mL of stock solution with 75 mL of water.

5. How many grams of NaOH would be needed to make 250 mL of 0.200 M solution? (molecular weight of NaOH = 40.0 g/mol)
   a. 8.00 g
   b. 4.00 g
   c. 2.00 g
   d. 2.50 g

6. The number of moles of NaCl in 250 mL in a 0.300 M solution of NaCl is
   a. 0.0750.
   b. 0.150.
   c. 0.250.
   d. 1.15.

7. A substance has the formula $MgSO_4 \cdot 7H_2O$. How many grams of water are in 5.00 moles of this substance?
   a. 7.00
   b. 35.0
   c. 126
   d. 630

8. How many grams of sugar are needed to make 500 mL of a 5% (weight/volume) solution of sugar?
   a. 20
   b. 25
   c. 50
   d. 10

9. What are the spectator ions in the following equation?

$Pb(NO_3)_2(aq) + 2KCl(aq) \rightarrow$

$PbCl_2(s) + 2KNO_3(aq)$

a. Pb^{2+}, $2NO_3^-$, $2K^+$, and $2Cl^-$
b. Pb^{2+} and $2NO_3^-$
c. Pb^{2+}, $2K^+$, and $2Cl^-$
d. $2NO_3^-$ and $2K^+$

10. What are the products of the following equation?

sodium chloride*(aq)* + lead(II) nitrate*(aq)* →

a. sodium nitrate + lead(II) chloride
b. sodium + chloride
c. sodium + chloride + lead(II) + nitrate
d. sodium(II) nitrate + lead chloride

11. Complete the following precipitation reaction:

$Hg_2(NO_3)_2(aq) + KI(aq) \rightarrow$

a. $Hg_2I_2(s) + 2K^+(aq) + 2NO_3^-(aq)$
b. $Hg_2I_2(s) + 2KNO_3(s)$
c. $Hg_2^{2+}(aq) + 2NO_3^-(aq) + 2K^+(aq) + 2I^-(aq)$
d. $Hg_2^{2+}(aq) + 2NO_3^-(aq) + 2KI(s)$

12. Which of the following is NOT true of reversible chemical reactions?

a. A chemical reaction is never complete.
b. The products of the reaction also react to reform the original reactants.
c. When the reaction is finished, both reactants and products are present in equal amounts.
d. The reaction can result in equilibrium.

13. Which is an example of an exothermic change?

a. sublimation
b. condensation
c. melting
d. evaporation

14. Which is an example of an endothermic change?

a. condensation
b. sublimation
c. freezing
d. combustion

15. The following reaction is exothermic: $AgNO_3 + NaCl \rightleftharpoons AgCl + NaNO_3$. How will the equilibrium be changed if the temperature is increased?

a. Equilibrium will shift to the right.
b. Equilibrium will shift to the left.
c. The reaction will not proceed.
d. Equilibrium will not change.

16. The pH of a blood sample is 7.40 at room temperature. The pOH is therefore

a. 6.60.
b. 7.40.
c. 6×10^{-6}.
d. 4×10^{-7}.

17. As the concentration of hydrogen ions in a solution decreases,

a. the pH numerically decreases.
b. the pH numerically increases.
c. the product of the concentrations $[H^+] \times [OH^-]$ comes closer to 1×10^{-14}.
d. the solution becomes more acidic.

18. The pH of an alkaline solution is

a. less than 0.
b. less than 7.
c. more than 14.
d. more than 7.

19. Which of the following is considered neutral on the pH scale?

a. pure water
b. pure saliva
c. pure blood
d. pure urine

20. A substance that functions to prevent rapid, drastic changes in the pH of a body fluid by changing strong acids and bases into weak acids and bases is called a(n)

a. salt.
b. buffer.
c. enzyme.
d. coenzyme.

21. Complete the following equation: $NaHCO_3$ + $HCl \rightarrow NaCl$ +

a. HCO_3
b. H_2CO_3
c. CO_2
d. H_2O

22. What is the oxidation number for nitrogen in HNO_3?

a. +2
b. +5
c. −2
d. −5

23. Identify the oxidizing agent and the reducing agent in the following reaction:

$$8NH_{3(g)} + 6NO_{2(g)} \rightarrow 7N_{2(g)} + 12H_2O_{(l)}$$

a. oxidizing agent $N_{2(g)}$, reducing agent $H_2O_{(l)}$
b. oxidizing agent $NH_{3(g)}$, reducing agent $NO_{2(g)}$
c. oxidizing agent $NO_{2(g)}$, reducing agent $N_{2(g)}$
d. oxidizing agent $NO_{2(g)}$, reducing agent $NH_{3(g)}$

24. Identify the oxidizing agent and the reducing agent in the following reaction:

$$8H^+(aq) + 6Cl^-(aq) + Sn(s) + 4NO_3^-(aq) \rightarrow SnCl_6^{2-}(aq) + 4NO_{2(g)} + 4H_2O_{(l)}$$

a. oxidizing agent $H^+(aq)$, reducing agent $Sn(s)$
b. oxidizing agent $NO_3^-(aq)$, reducing agent $Sn(s)$
c. oxidizing agent $NO_3^-(aq)$, reducing agent $NO_{2(g)}$
d. oxidizing agent $NO_3^-(aq)$, reducing agent $H^+(aq)$

25. Balance the following redox reaction:

$$Mg(s) + H_2O_{(g)} \rightarrow Mg(OH)_2(s) + H_{2(g)}$$

a. $Mg(s) + H_2O_{(g)} \rightarrow Mg(OH)_2(s) + H_{2(g)}$
b. $Mg(s) + 4H_2O_{(g)} \rightarrow Mg(OH)_2(s) + H_{2(g)}$
c. $Mg(s) + 2H_2O_{(g)} \rightarrow Mg(OH)_2(s) + H_{2(g)}$
d. $Mg(s) + H_2O_{(g)} \rightarrow Mg(OH)_2(s) + H_{2(g)}$

26. The time required for half the atoms in a sample of a radioactive element to disintegrate is known as the element's
a. decay period.
b. life time.
c. radioactive period.
d. half-life.

27. The half-life of a given element is 70 years. How long will it take 5.0 g of this element to be reduced to 1.25 g?
a. 70 years
b. 140 years
c. 210 years
d. 35 years

28. The elements found in carbohydrates are
a. oxygen, carbon, and hydrogen.
b. zinc, hydrogen, and iron.
c. carbon, iron, and oxygen.
d. hydrogen, iron, and carbon.

29. The complementary DNA base pair of thymine is
a. guanine
b. adenine
c. cytosine
d. uracil

30. The bonds between amino acids in a polypeptide are
a. glycosidic bonds.
b. ester bonds.
c. peptide bonds.
d. hydrogen bonds.

Answers

If you miss any of the answers, you can find help in the lesson shown to the right.

1. **b.** Lesson 6
2. **d.** Lesson 8
3. **d.** Lesson 8
4. **a.** Lesson 6
5. **c.** Lesson 6
6. **a.** Lesson 6
7. **d.** Lesson 4
8. **b.** Lesson 12
9. **d.** Lesson 12
10. **a.** Lesson 12
11. **a.** Lesson 12
12. **c.** Lesson 9
13. **b.** Lesson 10
14. **b.** Lesson 10
15. **c.** Lesson 9
16. **a.** Lesson 13
17. **b.** Lesson 13
18. **d.** Lesson 13
19. **a.** Lesson 13
20. **b.** Lesson 13
21. **b.** Lesson 12
22. **b.** Lesson 19
23. **d.** Lesson 19
24. **b.** Lesson 19
25. **c.** Lesson 19
26. **d.** Lesson 24
27. **b.** Lesson 24
28. **a.** Lesson 22
29. **c.** Lesson 23
30. **c.** Lesson 22

ANSWER KEY

Lesson 1

1. Form a hypothesis.
2. Research the problem.
3. Repeat the experiment
4. Form a conclusion.
5. Law
6. Theory
7. Law
8. Law
9. Metalloid
10. Nonmetal
11. Metal
12. Metal
13. Nonmetal

Lesson 2

1. $11p^+$, $12n^\circ$, $11e^-$
2. $43p^+$, $56n^\circ$, $43e^-$
3. $5p^+$, $6n^\circ$, $5e^-$
4. $15p^+$, $16n^\circ$, $15e^-$
5. $17p^+$, $18n^\circ$, $17e^-$
6. N^{3-}
7. K^+
8. I^-
9. Mg^{2+}
10. S^{2-}
11. Lead (II) oxide
12. Aluminum chloride (or aluminum trichloride)
13. Iron (III) oxide
14. Lithium fluoride
15. Zinc bromide
16. $HgCl_2$
17. $SrBr_2$
18. Na_2S
19. MgO
20. CoF_3
21. Dinitrogen tetraoxide
22. Diphosphorus pentoxide
23. Nitrogen monoxide
24. NO_3
25. ICl_3
26. CCl_4
27. Cesium perchlorate

28. Sodium bicarbonate (or sodium hydrogen carbonate)
29. Iron (III) nitrate
30. Hg_2Cl_2
31. $Cu(NO_3)_2$
32. CaC_2O_4
33. Perchloric acid
34. Hydrobromic acid
35. Hydrosulfuric acid
36. Phosphoric acid
37. Nitrous acid

Lesson 3

1. 5.974×10^{24} kg
2. 2.99×10^{8} m/s
3. 6.022×10^{23}
4. 9.11×10^{-31} kg
5. 4
6. 2
7. 4
8. 3
9. 5
10. 18
11. 5.02×10^{22}
12. -2.1
13. 10
14. 15.4
15. 5.6×10^{-5} km
16. 3.4×10^{-8} cm
17. 0.660 kg
18. 0.150 m
19. 14,600 mL
20. 64° F
21. 373 K
22. 29° C
23. 298 K
24. -38° C

Lesson 4

1. 62.01 g/mol
2. 60.05 g/mol
3. 1.09 moles
4. 150 g
5. 3.63×10^{23} molecules
6. 3.086% H, 31.61% P, 65.31% O
7. 52.14% C, 13.13% H, 34.73% O
8. 62.04% C, 10.41% H, 27.55% O
9. 69.94% Fe, 30.06% O
10. CH
11. N_2H_4CO
12. CH_4O
13. C_2H_3Cl

Lesson 5

1. $2C_2H_6 + 7O_2 \rightarrow 4CO_2 + 6H_2O$
2. $2Na + 2H_2O \rightarrow 2NaOH + H_2$
3. $NH_4NO_3 \rightarrow N_2O + 2H_2O$
4. $4NH_3 + 5O_2 \rightarrow 4NO + 6H_2O$
5. $Fe_2O_3 + 3C \rightarrow 2Fe + 3CO$
6. 124 g
7. 37 g
8. 3.38 g
9. Limiting reactant = H_2; theoretical yield = 8.4 g
10. Limiting reactant = O_2; theoretical yield = 12.7 g
11. Limiting reactant = NO_2; theoretical yield = 55 g
12. Limiting reactant = salicylic acid; theoretical yield = 62.8 g
13. 75%
14. 97.6%
15. 93%
16. 56%

Lesson 6

1. 4.5 M
2. 42.0 g
3. 2.2 M
4. 2.5 moles
5. 20.5 N
6. 0.21 moles
7. 0.10 L
8. 0.51 M
9. 0.078 L = 78 mL
10. 0.127 M
11. 75.0 mL
12. 50.0 mL

Lesson 7

1. van der Waals, dipole-dipole
2. van der Waals, dipole-dipole, ion-dipole
3. van der Waals, dipole-dipole, hydrogen bonding
4. van der Waals
5. van der Waals, dipole-dipole, ion-dipole, hydrogen bonding
6. van der Waals, dipole-dipole
7. van der Waals, dipole-dipole, hydrogen bonding
8. CH_3CH_2OH
9. SO_2
10. HF
11. Xe
12. Heterogeneous mixture
13. Homogeneous mixture
14. Element
15. Heterogeneous mixture
16. Compound
17. Homogeneous mixture
18. Heterogeneous mixture
19. Physical change
20. Chemical change
21. Physical change
22. Chemical change
23. Chemical change
24. 0.272 M
25. 4.45 M
26. 34.6° C
27. −3.23° C

Lesson 8

1. 0.737 atm
2. 125,000 Pa
3. 1,700 torr
4. 2.8 atm
5. 6.47 L
6. 37 psi
7. 11.4 moles
8. 0.103 g/L
9. 1.1 L
10. N_2 = 601 mm Hg; O_2 = 158 mm Hg
11. *Ptotal* = 12.2 atm (*poxygen* = 2.4, *pnitrogen* = 9.8)
12. 0.9357

Lesson 9

1. $K_{eq} = \dfrac{[NO]^4[H_2O]^6}{[NH_3]^4[O_2]^5}$
2. $K_{eq} = \dfrac{[HBr]^2}{[H_2][Br_2]}$
3. $K_{eq} = \dfrac{[NO_2]^4[O_2]}{[N_2O_5]^2}$
4. $K_{eq} = 1$
5. $K_{eq} = \dfrac{P_{SO_3}}{[H_2SO_4]}$
6. $K_{eq} = \dfrac{(PH_I)^2}{(PH_2)(PI_2)}$

7. 0.0480 M
8. 3.01×10^{-7} M
9. 0.38 M
10. Reaction shifts to the right.
11. Reaction shifts to the right.
12. Reaction shifts to the left.
13. Reaction shifts to the right.

Lesson 10

1. -2.02×10^6 J = $-2{,}020$ kJ
2. 8,010 J = 8.01 kJ
3. 33.6° C
4. $-1{,}428$ kJ, exothermic
5. 2,802 kJ, endothermic
6. -89 kJ, exothermic
7. v826 kJ, exothermic
8. $-1{,}774.0$ kJ
9. $+36$ kJ

Lesson 11

1. Rate = $k[N_2][H_2]^3$
2. Fourth order
3. One N_2 molecule is consumed for every two molecules of NH_3 produced.
4. Eight times faster
5. Rate = $k[H_2O_2]$
6. First order

Lesson 12

1. Strong electrolyte
2. Nonelectrolyte
3. Weak electrolyte
4. Strong electrolyte
5. Weak electrolyte
6. Soluble; Cr^{2+}, NO_3^-
7. Insoluble
8. Soluble; Ba^{2+}, OH^-
9. Insoluble
10. Soluble; K^+, PO_4^{3-}
11. $NH_4Cl(aq) + AgNO_3(aq) \rightarrow NH_4NO_3(aq) + AgCl(s)$; $Cl^-(aq) + Ag^+(aq) \rightarrow AgCl(s)$
12. $2NaOH(aq) + MgCl_2(aq) \rightarrow 2NaCl(aq) + Mg(OH)_2(s)$; $2OH^-(aq) + Mg^{2+}(aq) \rightarrow Mg(OH)_2(s)$
13. $Pb(C_2H_3O_2)_2(aq) + Na_2SO_4(aq) \rightarrow PbSO_4(s) + 2NaC_2H_3O_2(aq)$; $Pb^{2+}(aq) + SO_4^{2-}(aq) \rightarrow PbSO_4(s)$
14. $KCl(aq) + Li_2CO_3(aq) \rightarrow$ no reaction
15. $K_2S(aq) + Ni(NO_3)_2(aq) \rightarrow 2KNO_3(aq) + NiS(s)$; $S^{2-}(aq) + Ni^{2+}(aq) \rightarrow NiS(s)$
16. Acid
17. Acid
18. Base
19. Acid
20. Base
21. $HCN(aq) + OH-(aq) \rightarrow CN-(aq) + H_2O(l)$
22. $Al(OH)_3(s) + 3H^+(aq) \rightarrow Al^{3+}(aq) + 3H_2O(l)$
23. $CH_3CO_2H(aq) + OH^-(aq) \rightarrow CH_3CO_2^-(aq) + H_2O(l)$
24. $OH^-(aq) + H^+(aq) \rightarrow H_2O(l)$
25. $2HNO_2(aq) + Mg(OH)_2(s) \rightarrow Mg(NO_2)_2(s) + 2H_2O(l)$
26. 0
27. K = +1, N = +5, O = −2
28. H = +1, Cr = +6, O = −2
29. H = +1, O = −1
30. P = +3, F = −1
31. Species being oxidized = Mg, species being reduced = H, oxidizing agent = H_2O, reducing agent = Mg
32. Species being oxidized = Sn, species being reduced = N, oxidizing agent = NO_3^-, reducing agent = Sn

33. Species being oxidized = Fe, species being reduced = Mn, oxidizing agent = MnO_4^-, reducing agent = Fe^{2+}

34. Species being oxidized = Na, species being reduced = Cl, oxidizing agent = Cl_2, reducing agent = Na

35. Species being oxidized = H, species being reduced = O, oxidizing agent = O_2, reducing agent = H_2

Lesson 13

1. $H_2SO_4 + 2NaOH \rightarrow Na_2SO_4 + 2H_2O$
2. $2Al + 6HCl \rightarrow 2AlCl_3 + 3H_2$
3. $Li_2O + 2HNO_3 \rightarrow 2LiNO_3 + H_2O$
4. $CaCO_3 + 2HCl \rightarrow CaCl_2 + CO_2 + H_2O$
5. 2.3
6. 12 (pOH = 2)
7. −1.08
8. 1.90
9. 2.87
10. 2.33
11. 4.36
12. 6.84
13. 6.00

Lesson 14

1. 2.00×10^3 nm
2. 2.21×10^{-21} J
3. 4.6×10^{14} Hz
4. 1.45×10^{12} Hz
5. $E = 2.09 \times 10^{-18}$ J, $\lambda = 94.9$ nm
6. $E = -4.09 \times 10^{-19}$ J, $\lambda = 486$ nm
7. $E = -2.15 \times 10^{-18}$ J, $\lambda = 92.6$ nm
8. $E = -5.45 \times 10^{-19}$ J, $\lambda = 364$ nm

Lesson 15

1. no, $n \neq 0$
2. no, m_l must be −1, 0, or 1 when $l = 1$
3. yes
4. yes
5. no, m_s can only be $+\frac{1}{2}$ or $-\frac{1}{2}$
6. [Kr] $5s^2 4d^{10}$
7. [Ar] $4s^2 4p^6$
8. [Kr] $5s^2 4d^1$
9. [Ar]
10. [Ne] $3s^2 3p^3$
11. Si:

 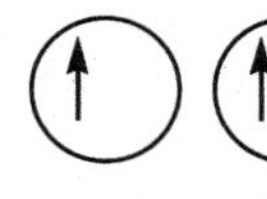 

3s 3p

12. Pd:

  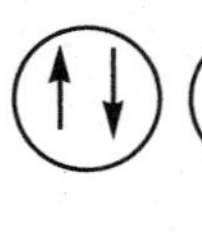 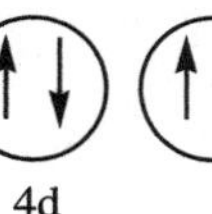 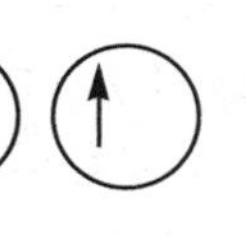

5s 4d

13. N^{3-}:

   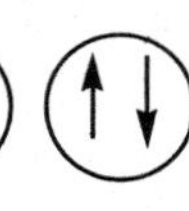

2s 2p

14. B:

 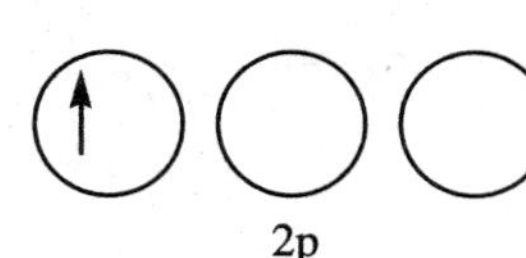

2s 2p

15. V^{5+}:

3s 3p

Lesson 16

1.

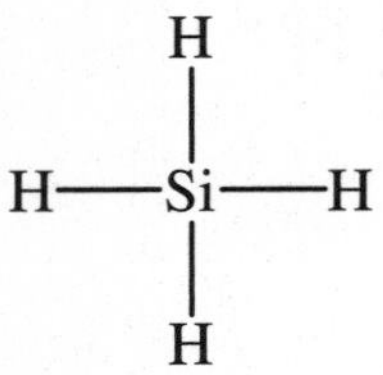

H
H—Si—H
H

2.

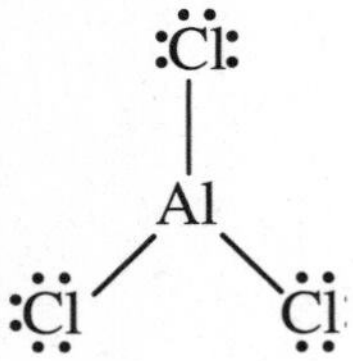

:Cl:
Al
:Cl Cl:

3.

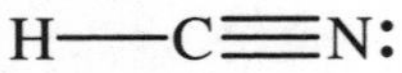

H—C≡N:

4.

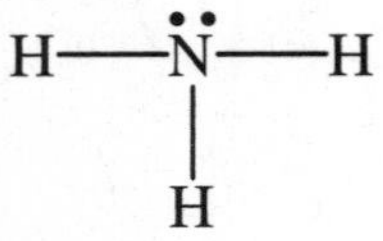

H—N—H
H

5.

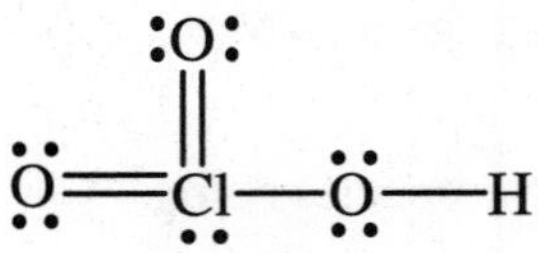

:O:
O=Cl—O—H

6.

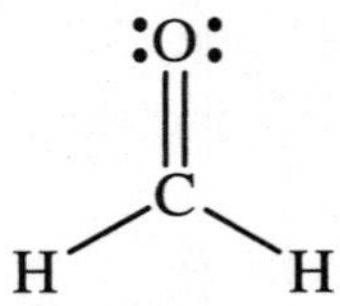

:O:
C
H H

7.

:F:
:F—Xe—F:
:F:

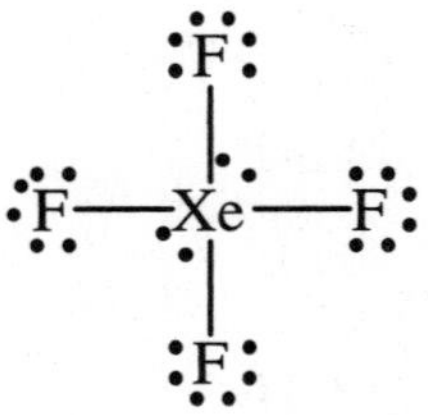

8. Trigonal planar
9. Bent
10. Trigonal planar
11. Tetrahedral
12. Trigonal bipyramidal
13. Seesaw

14.

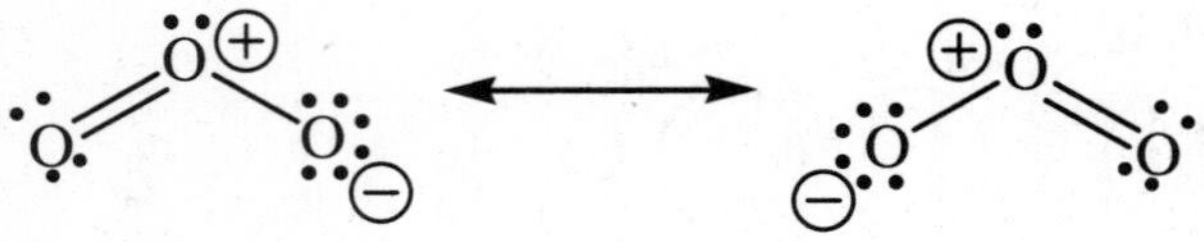

15.

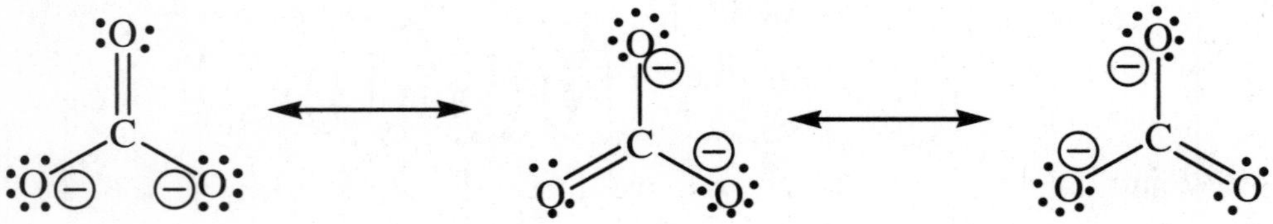

16.

17.

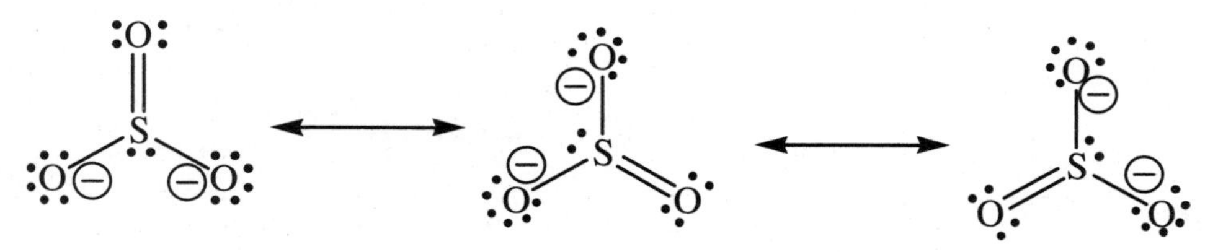

18.

19.

20.

21.

22.

H—Si

23.

C—F

24.

B—Cl

25.

S—F

26. Net dipole exists

27. No net dipole

28. Net dipole exists

Lesson 17

1. 4 bonding, 4 antibonding, BO = 0.
2. 10 bonding, 8 antibonding, BO = 1.
3. He^{2+}: BO = ½

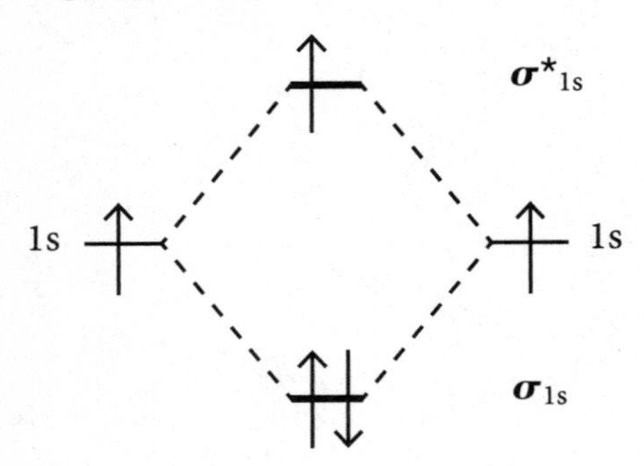

4. BO = 1, compound should exist
5. BO = $\frac{(6 - 2)}{2 = 2}$

Lesson 18

1. Diamagnetic
2. It is diamagnetic.
3. Paramagnetism
4. Li and B
5. Ferromagnetism

Lesson 19

1. $8H + 3Cu + 2NO_3^- \rightarrow 3Cu^{2+} + 2NO + 4H_2O$
2. $8H^+ + MnO_4^- + 5Fe^{2+} \rightarrow 5Fe^{3+} + Mn^{2+} + 4H_2O$
3. $H^+ + Zn + HCl \rightarrow Zn^{2+} + H_2 + Cl^-$
4. $2OH^- + Cl_2 \rightarrow ClO^- + Cl^- + H_2O$
5. $H_2O_2 + SO_3^{2-} \rightarrow H_2O + SO_4^{2-}$
6. $2Fe^{2+} + H_2O_2 \rightarrow 2Fe^{3+} + 2OH^-$

Lesson 20

1. 2-methylhexane
2. 2,3-dimethylbutane
3. 4-ethyl-3-methylheptane

4 & 5.

$CH_3CH_2CH_2CH_2CH_2CH_3$
hexane

$CH_3CH_2CH_2CH(CH_3)CH_3$
2-methylhexane

$CH_3CH_2CH(CH_3)CH_2CH_3$
3-methylhexane

$CH_3C(CH_3)_2CH_2CH_3$
2,2-dimethylbutane

$CH_3CH(CH_3)CH(CH_3)CH_3$
2,3-dimethylbutane

6. 3-heptyne
7. *cis*-2-petene
8. *trans*-3-hexene
9.

$H_3CH_2C(H)C{=}C(H)(CH_2)_5CH_3$

10.

$CH_3CH_2C{\equiv}CCH_3$

11.

$H_3C(H)C{=}C(H)(CH_2)_4CH_3$

Lesson 21

1. Ethanol
2. Pentanol
3. Hexanol
4. Methanol
5. Diethylether
6. Methylethylether
7. Dipropylether
8. Dimethylether
9. Formaldehyde
10. Propylaldehyde
11. Acetaldehyde
12. Butylaldehyde
13. Dimethylketone
14. Ethylpropylketone
15. Methylethylketone
16. Methylbutylketone
17. Propanoic acid
18. Acetic acid
19. Butanoic acid
20. Formic acid
21. Methyl acetate
22. Ethyl butyrate
23. Ethyl formate

24.

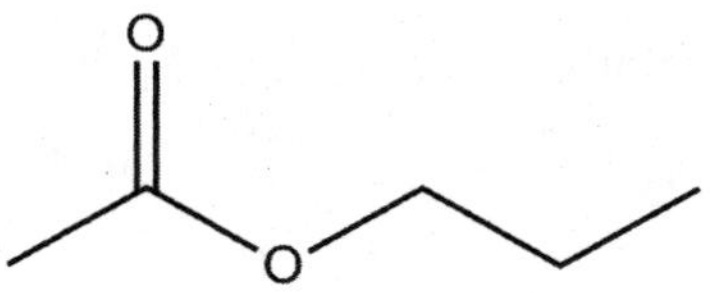

25.

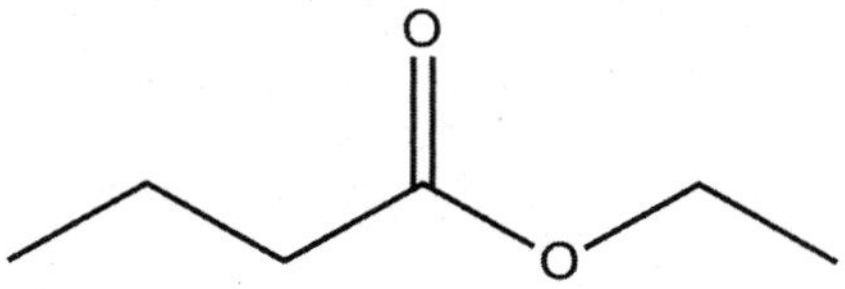

26.

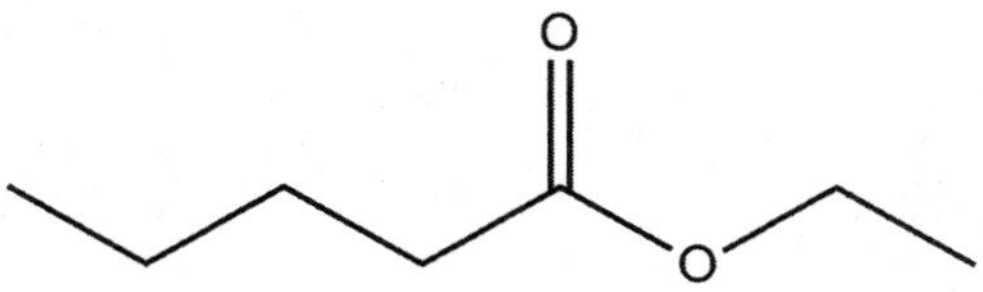

27. Trimethylamine, 3°
28. Ethylamine 1°
29. Dimethylethylamine, 3°
30. Dipropylamine, 2°
31.

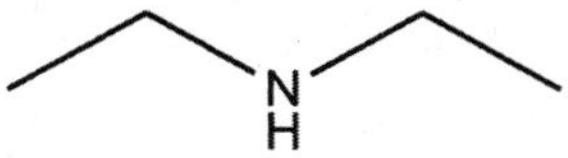

32.

33.

34.

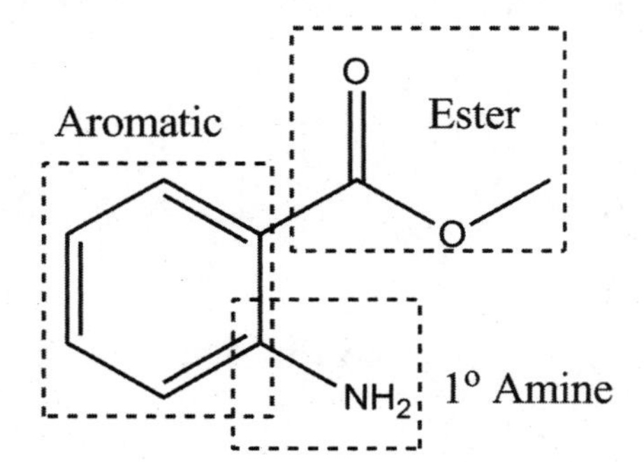

35.

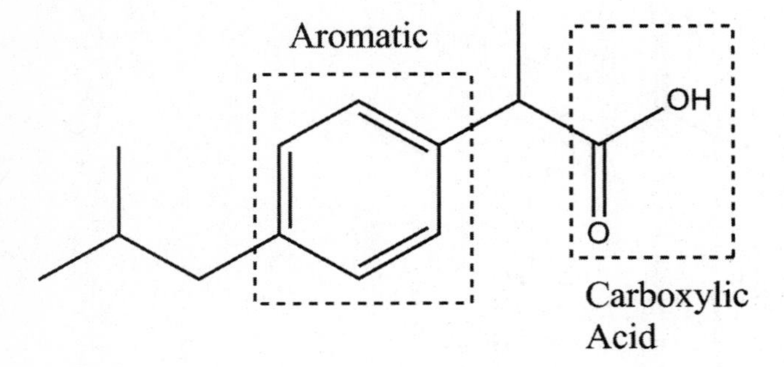
Aromatic
OH
O
Carboxylic
Acid

36.

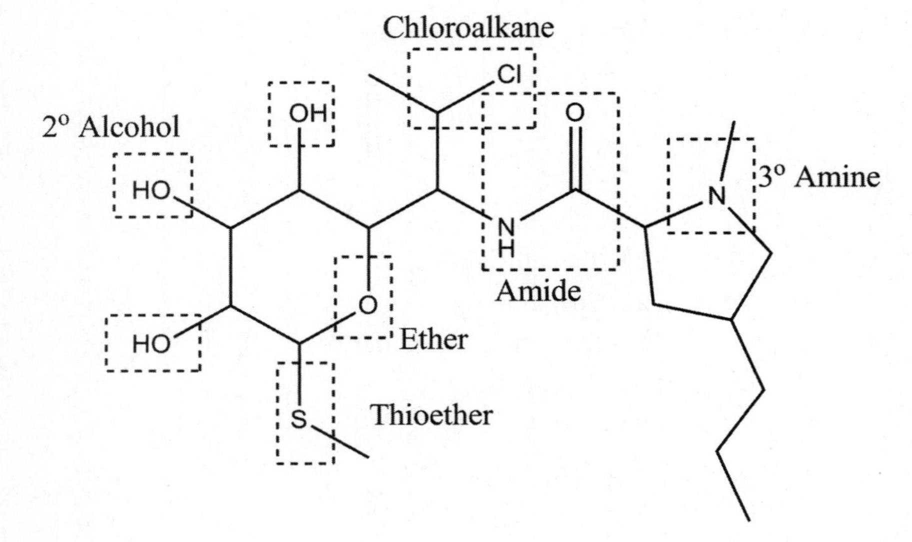
Chloroalkane
Cl
O
2° Alcohol
OH
HO
N
3° Amine
N
H
Amide
O
HO
Ether
S
Thioether

37.

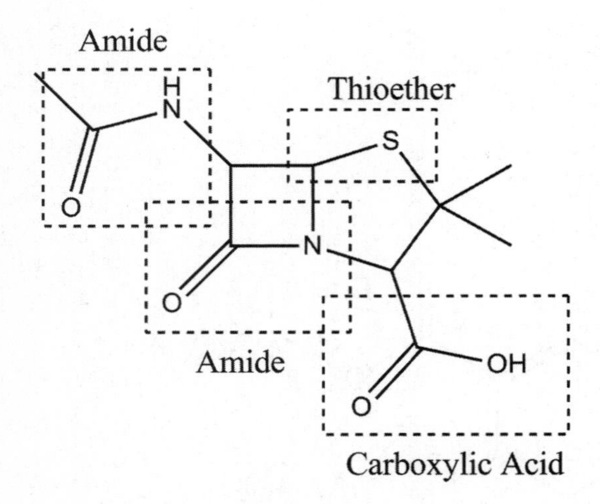
Amide
H
N
Thioether
S
O
N
O
Amide
OH
O
Carboxylic Acid

Lesson 22

1. Carbon, hydrogen, oxygen
2. Energy
3. Lipids
4. Fats
5. D-glucose, D-fructose
6. Peptide bonds
7. Glycogen
8. Active site

Lesson 23

1. TACGAGTTGCTCGCTATT
2. TACCTTGAAGTAAAAATC
3. UCAUUGGUC
4. Start-Leu-Asn-Glu-Arg-Stop
5. Start-Glu-Leu-His-Phe-Stop

Lesson 24

1. α particles
2. ${}^{238}_{93}\mathrm{U}$
3. ${}^{0}_{-1}\beta$
4. ${}^{4}_{2}\mathrm{He}$
5. ${}^{4}_{2}\mathrm{He}$
6. ${}^{0}_{-1}\mathrm{e}$
7. Fission and fusion
8. 28.8 hours
9. 8.10 days
10. 42.9 days

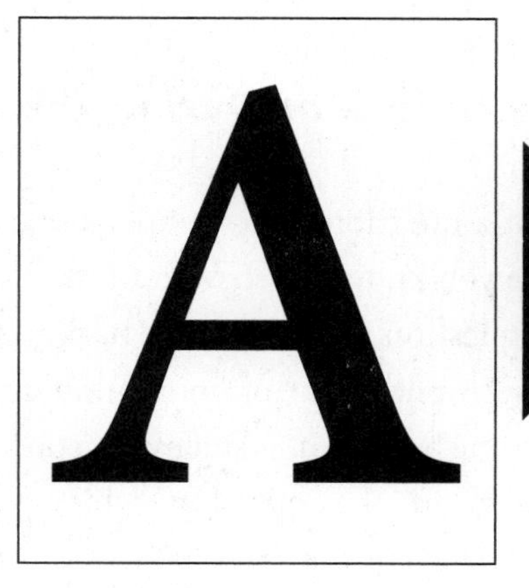

How to Prepare For a Test

A standardized test is nothing to fear. Many people worry about a testing situation, but you're much better off taking that nervous energy and turning it into something positive that will help you do well on your test, rather than inhibit your testing ability. The following pages include valuable tips for combating test anxiety—that sinking or blank feeling some people get as they begin a test or encounter a difficult question. You will also find valuable tips for using your time wisely and for avoiding errors in a testing situation. Finally, you will find a plan for preparing for the test, a plan for the test day, and a suggestion for a posttest activity.

Combating Test Anxiety

Knowing what to expect and being prepared are the best defense against test anxiety—that worrisome feeling that keeps you from doing your best. Practice and preparation keep you from succumbing to that anxiety. Nevertheless, even the brightest, most well-prepared test takers may suffer from occasional bouts of test anxiety. But don't worry; you can overcome it.

Take the Test One Question at a Time

Focus all your attention on the one question you're answering. Block out any thoughts about questions you've already read or concerns about what is coming next. Concentrate your thinking where it will do the most good—on the question you're answering.

Develop a Positive Attitude

Keep reminding yourself that you're prepared. The fact that you have read this book means that you're better prepared than most taking the test. Remember, it's only a test, and you're going to do your best. That's all anyone can ask of you. If that nagging drill sergeant voice inside your head starts sending negative messages, fight back with positive ones of your own:

- "I'm doing just fine."
- "I've prepared for this test."
- "I know exactly what to do."
- "I know I can get the score I'm shooting for."

You get the idea. Remember to drown out negative messages with positive ones of your own.

If You Lose Your Concentration

Don't worry about it! It's normal. During a long test, it happens to everyone. When your mind is stressed or overexerted, it takes a break whether you want it to or not. It's easy to get your concentration back if you simply acknowledge the fact that you've lost it and take a quick break. You brain needs very little time (seconds really) to rest.

Put your pencil down and close your eyes. Take a few deep breaths and listen to the sound of your breathing. The ten seconds or so that this takes is really all the time your brain needs to relax and get ready to focus again.

Try this technique several times in the days before the test when you feel stressed. The more you practice, the better it will work for you on the day of the test.

If You Freeze Before or During the Test

Don't worry about a question that stumps you, even though you're sure you know the answer. Mark it and go on to the next question. You can come back to the stumper later. Try to put it out of your mind completely until you come back to it. Just let your subconscious mind chew on the question while your conscious mind focuses on the other items (one at a time, of course). Chances are, the memory block will be gone by the time you return to the question.

If you freeze before you begin the test, here's what to do:

1. Take a little time to look over the test.
2. Read a few of the questions.
3. Decide which ones are the easiest and start there.
4. Before long, you'll be in the groove.

Time Strategies

Use your time wisely to avoid making careless errors.

Pace Yourself

The most important time strategy is pacing yourself. Before you begin, take just a few seconds to survey the test, noting the number of questions and the sections that look easier than the rest. Plan a rough time schedule based on the amount of time available to you. Mark the halfway point on your test and indicate what the time will be when the testing period is half over.

Keep Moving

Once you begin the test, keep moving. If you work slowly in an attempt to make fewer mistakes, your mind will become bored and begin to wander. You'll end up making far more mistakes if you're not concentrating.

As long as we're talking about mistakes, don't stop for difficult questions. Skip them and move on. You can come back to them later if you have time. A question that takes you five seconds to answer counts as much as one that takes you several minutes, so pick up the easy points first. Besides, answering the easier questions first helps build your confidence and gets you in the testing groove. Who knows? As you go through the test, you may even stumble across some relevant information to help you answer those tough questions.

Don't Rush

Keep moving, but don't rush. Think of your mind as a seesaw. On one side is your emotional energy; on the other side, your intellectual energy. When your emotional energy is high, your intellectual capacity is low. Remember how difficult it is to reason with someone when you're angry? On the other hand, when your intellectual energy is high, your emotional energy is low. Rushing raises your emotional energy. Remember the last time you were late for work? All that rushing around causes you to forget important things, such as your lunch. Move quickly to keep your mind from wandering, but don't rush and get yourself flustered.

Check Yourself

Check yourself at the halfway mark. If you're a little ahead, you know you're on track and may even have a little time left to check your work. If you're a little behind, you have several choices. You can pick up the pace a little, but do this only if you can do it comfortably. Remember—don't rush! You can also skip around in the remaining portion of the test to pick up as many easy points as possible. This strategy has one drawback, however. If you are marking a bubble-style answer sheet, and you put the right answers in the wrong bubbles, they're wrong. So pay close attention to the question numbers if you decide to do this.

Avoiding Errors

When you take the test, you want to make as few errors as possible in the questions you answer. Here are a few tactics to keep in mind.

Control Yourself

Remember the comparison between your mind and a seesaw that you read about a few paragraphs ago? Keeping your emotional energy low and your intellectual energy high is the best way to avoid mistakes. If you feel stressed or worried, stop for a few seconds. Acknowledge the feeling (Hmmm! I'm feeling a little pressure here!), take a few deep breaths, and send yourself a few positive messages. This relieves your emotional anxiety and boosts your intellectual capacity.

Directions

In many standardized testing situations, a proctor reads the instructions aloud. Make certain you understand what is expected. If you don't, ask. Listen carefully for instructions about how to answer the questions and make certain you know how much time you have to complete the task. Write the time on your test if you don't already know how long you have. If you miss this vital information, ask for it. You need it to do well.

Answers

Place your answers in the right blanks or in the corresponding ovals on the answer sheet. Right answers in the wrong place earn no points. It's a good idea to check every five to ten questions to make sure you're in the right spot. That way you won't need much time to correct your answer sheet if you have made an error.

Reading Long Passages

Frequently, standardized tests are designed to test your reading comprehension. The reading sections often contain passages that are a paragraph or more. Here are a few tactics for approaching these sections.

This may seem strange, but some questions can be answered without ever reading the passage. If the passage is short, a paragraph around four sentences or so, read the questions first. You may be able to answer them by using your common sense. You can check your answers later after you've actually read the passage. Even if you can't answer any of the questions, you know what to look for in the passage. This focuses your reading and makes it easier for you to retain important information. Most questions will deal with isolated details in the passage. If you know what to look for ahead of time, it's easier to find the information.

If a reading passage is long and followed by more than ten questions, you may end up spending too much time reading the questions first. Even so, take a few seconds to skim the questions and read a few of the shorter ones. As you read, mark up the passage. If you find a sentence that seems to state the main idea of the passage, underline it. As you read through the rest of the passage, number the main points that support the main idea. Several questions will deal with this information and if it's underlined and numbered, you can locate it easily. Other questions will ask for specific details. Circle information that tells who, what, when, or where. The circles will be easy to locate later if you run across a question that asks for specific information. Marking up a passage this way also heightens your concentration and makes it more likely that you'll remember the information when you answer the questions following the passage.

Choosing the Right Answers

Make sure you understand what the question is asking. If you're not sure what's being asked, you'll never know whether you've chosen the right answer, so figure out what the question is asking. If the answer isn't readily apparent, look for clues in the answer choices. Notice the similarities and differences. Sometimes this helps put the question in a new perspective and makes it easier to answer. If you're still not sure of the answer, use the process of elimination. First, eliminate any answer choices that are obviously wrong. Then, reason your way through the remaining choices. You may be able to use relevant information from other parts of the test. If you can't eliminate any of the answer choices, you might be better off to skip the question and come back to it later. If you can't eliminate any answer choices to improve your odds when you come back later, then make a guess and move on.

If You're Penalized for Wrong Answers

You must know whether wrong answers are penalized before you begin the test. If you don't, ask the proctor before the test begins. Whether you make a guess depends on the penalty. Some standardized tests are scored in such a way that every wrong answer reduces your score by one-fourth or one-half a point. Whatever the penalty, if you can eliminate enough choices to make the odds of answering the question better than the penalty for getting it wrong, make a guess.

Let's imagine you are taking a test in which each answer has four choices and you are penalized one-fourth a point for each wrong answer. If you have no clue and cannot eliminate any of the answer choices, you're better off leaving the question blank because the odds of answering correctly are one in four. This makes the penalty and the odds equal. However, if you can eliminate one of the choices, the odds are now in your favor. You have a one in three chance of answering the question correctly. Fortunately, few tests are scored using such elaborate means, but if your test is one of them, know the penalties and calculate your odds before you take a guess on a question.

If You Finish Early

Use any time you have left at the end of the test or test section to check your work. First, make certain you've put the answers in the right places. As you're doing this, make sure you've answered each question only once. Most standardized tests are scored in such a way that questions with more than one answer are marked wrong. If you've erased an answer, make sure you've done a good job. Also, check for stray marks on your answer sheet that could distort your score.

After you've checked for these obvious errors, take a second look at the more difficult questions. You've probably heard the folk wisdom about never changing an answer. If you have a good reason for thinking a response is wrong, change it.

The Days before the Test

To get ready for a challenge like a big exam, you have to take control of your physical state as well as your mental state. Exercise, proper diet, and rest will ensure that your body works with—rather than against—your mind on test day, as well as during your preparation.

Physical Activity

Get some exercise in the days preceding the test. You'll send some extra oxygen to your brain and allow your thinking performance to peak on the day you take the test. Moderation is the key here. You don't want to exercise so much that you feel exhausted, but a little physical activity will invigorate your body and brain.

Balanced Diet

Like your body, your brain needs the proper nutrients to function well. Eat plenty of fruits and vegetables in the days before the test. Foods that are high in lecithin, such as fish and beans, are especially good choices. Lecithin is a mineral your brain needs for peak performance. You may even consider a visit to your local pharmacy to buy a bottle of lecithin tablets several weeks before your test.

Rest

Get plenty of sleep the nights before you take the test. Don't overdo it, though, or you'll make yourself as groggy as if you were overtired. Go to bed at a reasonable time, early enough to get the number of hours you need to function effectively. You'll feel relaxed and rested if you've gotten plenty of sleep in the days before you take the test.

Trial Run

At some point before you take the test, make a trial run to the testing center to see how long it takes. Rushing raises your emotional energy and lowers your intellectual capacity, so you want to allow plenty of time on test day to get to the testing center. Arriving 10 or 15 minutes early gives you time to relax and get situated.

Test Day

It's finally here, the day of the big test. Set your alarm early enough to allow plenty of time. Eat a good breakfast and avoid anything that's really high in sugar, such as donuts. A sugar high turns into a sugar low after an hour or so. Cereal and toast or anything with complex carbohydrates is a good choice. Eat only moderate amounts. You don't want to take a test feeling stuffed!

Pack a high-energy snack to take with you because you may have a break sometime during the test when you can grab a quick snack. Bananas are great. They have a moderate amount of sugar and plenty of brain nutrients such as potassium. Most proctors won't allow you to eat a snack while you're testing, but a peppermint shouldn't be a problem. Peppermints are like smelling salts for your brain. If you lose your concentration or suffer from a momentary mental block, a peppermint can get you back on track. Don't forget the earlier advice about relaxing and taking a few deep breaths.

Leave early enough so you have plenty of time to get to the test center. Allow a few minutes for unexpected traffic. When you arrive, locate the restroom and use it. Few things interfere with concentration as much as a full bladder. Then, find your seat and make sure it's comfortable. If it isn't, tell the proctor and ask to find a more suitable seat.

Now relax and think positively! Before you know it, the test will be over, and you'll walk away knowing you've done as well as you can.

After the Test

Do two things:

1. Plan a celebration.
2. Go to it.

If you have something to look forward to after the test is over, you may find it easier to prepare well for the exam and keep moving during the test. Good luck!

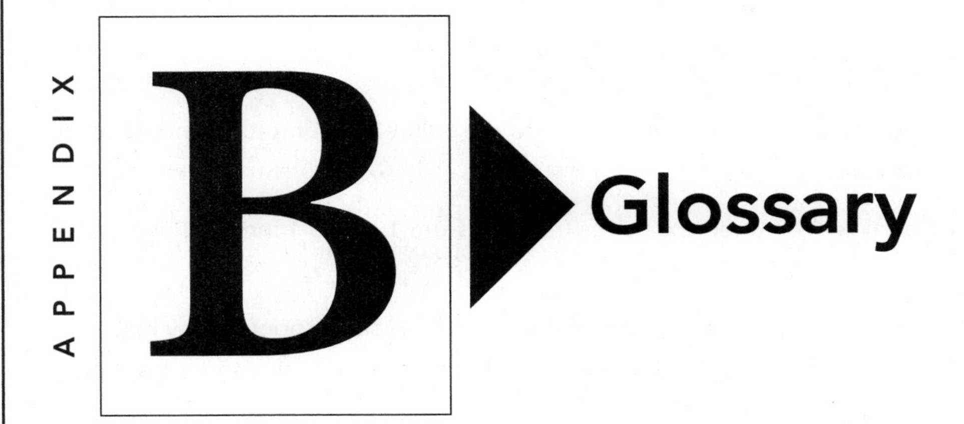

Glossary

Accuracy How close in agreement a measurement is with the accepted value.

Acid Substances that donate positive hydrogen ions (H^+) when dissolved in water.

Angular momentum quantum number The subshell designation of an electron that describes the shape of the orbital. The possible values for a particular energy level are 0 to ($n-1$).

Atom The basic unit of an element that retains all the element's chemical properties. An atom is composed of a nucleus (which contains one or more protons and neutrons) and one or more electrons in motion around it. Atoms are electrically neutral because they are made of an equal number of protons and electrons.

Atomic mass (weight) The weighted average of the isotopes' masses.

Atomic number The number of protons in the atom. The atomic number defines the element.

Aufbau Principle The Aufbau or "building-up" principle is based on the Pauli exclusion principle and states that electrons are placed in the most stable orbital.

Avogadro's Law The volume of gas increases with the number of moles of gas present at constant temperature and pressure ($V \alpha n$).

Avogadro's number $N_A = 6.022 \times 10^{23}$ items/mole.

Boyle's Law The volume of a gas (maintained at constant temperature) decreases as its pressure increases ($P \alpha 1/V$).

Carbohydrates (Or sugars) have the formula $C_x(H_2O)_y$ and serve as the main source of energy for living organisms.

Charles's Law The volume of a gas (maintained at constant pressure) increases directly with an increase in its Kelvin temperature ($V \alpha T$).

Chemical change A process where one or more substances are converted into one or more new substances.

Chemical equation A description of a chemical change or chemical reaction.

Colligative properties Solution properties that vary in proportion to the solute concentration and depend *only* on the number of solute particles.

Colloids Stable mixtures in which particles of rather large sizes (ranging from 1 nm to 1 mm) are dispersed throughout another substance.

Compound A combination of two or more atoms of different elements in a precise proportion by mass. In a compound, atoms are held together by attractive forces called chemical bonds and cannot be separated by physical means.

Diamagnetic Elements with all paired electrons and do not possess the ability to attract to a magnetic field.

Diffusion The mixing of gases.

Effusion The passage of a gas through a tiny hole usually into a chamber of lower pressure.

Electrolyte A solution that has electrical conductivity properties.

Electron A particle that is of negligible mass (0.000549 amu) compared to the mass of the nucleus. It has an effective negative charge of -1.

Electron spin quantum number The electron spin quantum number describes the spin of an electron.

Element A substance that contains one type of atom and cannot be broken down by simple means.

Empirical formula The simplest whole-number ratio of atoms in a molecule.

Endothermic Reactions or processes that consume energy.

Enzymes Biological catalysts whose role is to increase the rate of chemical (metabolic) reactions without being consumed in the reaction.

Equilibrium When two opposing reactions occur at the same rate.

Error The difference between a value obtained experimentally and the standard value accepted by the scientific community.

Exothermic Energy-releasing reactions or processes.

First law of thermodynamics Energy is neither created nor destroyed.

Formal charge The difference in the number of valance electrons in the neutral atom (group number) and the number of electrons assigned to that atom in the molecule or polyatomic ion.

Free energy (G) The energy state function of a system where $\Delta G = \Delta H - T\Delta S$

Gay-Lussac's Law The pressure of a gas (maintained at constant volume) increases with an increase in its Kelvin temperature ($P \alpha T$)

Half-life The time required for the concentration of the nuclei in a given sample to decrease to half its initial concentration.

Heat The transfer of thermal energy.

Heat capacity The amount of energy required to raise the temperature of a substance by 1° C.

Heat of fusion The heat required to fuse or melt a substance.

Heat of vaporization The heat required to evaporate 1 g of a liquid.

Henry's Law The solubility of an ideal gas (C) in moles per liter is directly proportional to the partial pressure (*p*) of the gas relative to a known constant (*k*) for the solvent and gas at a given temperature ($C = kp$).

Hess's Law An enthalpy of a reaction can be calculated from the sum of two or more reactions.

Heterogeneous mixture A system of two or more substances (elements or compounds) that have distinct chemical and physical properties. Examples include mixtures of salt and sand, oil and water, crackerjacks, and dirt.

Homogeneous mixture (or solution) A system of two or more substances (elements or compounds) that are interspersed like the gases making up the air or salt dissolved in water. The individual substances have distinct chemical properties and can be separated by physical means.

Hund's Rule The most stable arrangement of electrons in the same energy level is the one in which electrons have parallel spins (same orientation).

Ideal gas law $PV = nrt$.

Ionic equation Shows the strong electrolytes (soluble compounds as predicted by the solubility rules) as ions.

Isoelectric Having the same number of electrons.

Isomers Molecules with the same molecular formula, but a different structural arrangement.

Isotopes Atoms of an element that have different masses.

Kinetic molecular theory Describes the behavior of gases.

Law A law explains *what* happens.

Law of conservation of mass This states that mass cannot be created nor destroyed.

Law of definite proportions Different samples of the same compound always contain the same proportion by mass of each element.

Law of multiple proportions If two elements combine to form multiple compounds, the ration of the mass of one element combined with 1 gram of the other element can always be reduced to a whole number.

Le Châtelier's Principle If an equilibrium system is stressed, the equilibrium will shift in the direction that relieves the stress.

Limiting reactant or reagent The reagent that is consumed first in a reaction.

Lipids A diverse group of compounds that are insoluble in water and polar solvents but soluble in nonpolar solvents.

Magnetic quantum number The magnetic quantum number (m_l) describes the orbital's orientation in space.

Mass number The sum of protons and neutrons (in the nucleus) of an atom.

Matter Anything that occupies space and has mass; matter is everything in the universe.

Metal An element that is shiny, conducts electricity and heat, is malleable (easily shaped), and is ductile (pulled into wires).

Metalloid (or semimetal) An element with properties that are intermediate between those of metals and nonmetals such as semiconductivity.

Model The description of the theory.

Molality (m) The number of moles of a solute per kilogram of solvent.

Molarity (M) The number of moles of solute per liter of solution.

Mole A mole of a particular substance is equal to the number of atoms in exactly 12 g of the carbon-12 isotope. *See also* Avagadro's number.

Molecule A combination of two or more atoms. Molecules cannot be separated by physical means.

Molecular equation Shows the reactants and products as molecules.

Net ionic equation Shows only the species that are directly involved in the reaction (i.e., the spectator ions are not included).

Neutron A particle that has a mass of 1 atomic mass unit (amu; 1 amu = 1.66×10^{-27} kg) with no charge.

Node Represents an area where there is zero probability of an electron.

Nonmetal An element with poor conducting properties, is usually electronegative, and has a greater tendency to gain valence electrons.

Normality (N) The number of equivalents of the solute per liter of solution.

Nuclear fission The process in which a heavier nucleus (usually less stable) splits into smaller nuclei and neutrons.

Nuclear fusion The process in which small nuclei are combined (i.e., fused) into larger (more stable) ones with the release of a large amount of energy.

Orbital The space where one or two paired electrons *can* be located or the probability of an electron's location.

Oxidation A loss of electrons.

Oxidation number *See* oxidation state.

Oxidation state The number of charges carried by an ion or that an atom would have in a (neutral) molecule if electrons were transferred completely.

Oxidizing agent Contains the species being reduced and *helps* other compounds to be oxidized, hence being reduced itself.

Paramagnetic Elements that are attracted to a magnet.

Pauli Exclusion Principle States that an orbital can hold a maximum of two electrons if they are of opposite spins.

Physical change A physical change of a substance does not change its chemical composition.

Polyatomic ions Ions that contain more than one atom.

Polyprotic acids Substances containing more than one acidic proton.

Precipitation reaction When two soluble compounds are mixed and produce one or more insoluble compounds.

Precision The degree to which successive measurements agree with each other.

Pressure The force exerted over a unit area.

Principal quantum number The energy level of the electron. The value of n can be any integer.

Principle An explanation of a more specific set of relationships of a law.

Products A substance that is formed as a result of a chemical reaction (the right side of the arrow of a chemical equation).

Proteins Proteins (also called polypeptides) are long chains of amino acids joined together by covalent bonds of the same type (peptide or amide bonds).

Proton A particle that has a mass of 1 atomic mass unit (amu; 1 amu $= 1.66 \times 10^{-27}$ kg) and an effective positive charge of $+1$.

Raoult's Law (liquid-liquid solution) The vapor pressure of an ideal solution of 2 liquids (p_{total}) is directly proportional to the vapor pressures (p_A° and p_B°) of the pure liquids, the mole fractions of the liquids (X_A and X_B), and partial vapor pressure (p_a and p_b) of the liquids above the solution.

Raoult's Law (solid-liquid solution) The vapor pressure of an ideal solution (p_{total}) is directly proportional to the partial vapor pressure (p_a) of the pure solvent times the mole fraction (X_A) of the solute ($p_{total} = x_a p_a$).

Reactants A substance that undergoes a change in a chemical reaction (the left side of the arrow of a chemical equation).

Reducing agent Contains the species being oxidized and *helps* other compounds to be reduced, hence being oxidized itself.

Reduction A gain of electrons.

Resonance Occurs when one or more valid Lewis structures exist for a molecule or polyatomic ion.

Scientific method A framework for the stepwise process to experimentation.

Second law of thermodynamics There is an increase in entropy (randomness) in a spontaneous process.

Semimetal *See* metalloid.

SI units SI units (Système Internationale d'Unités) are the base units that are used by the modern metric system. Solute a substance dissolved in a solvent, forming a solution.

Solution *See* homogeneous mixture.

Spectator ions Ions not involved in the reaction.

Standard enthalpy of formation (DH°f) The energy required to form *one* mole of a substance from its elements in their standard states.

Stereoisomer Stereoisomers are isomers with the same connectivity, but a different three-dimensional structure.

Stoichiometry The quantities of reactants (used) and products (obtained) based on a balanced chemical equation.

Temperature The measure of thermal energy (total energy of all the atoms and molecules) of a system.

Theory A theory explains *why* something happens.

Thermodynamics The study of energy and its processes.

Triglycerides Triglycerides are lipids formed by condensation of glycerol (one molecule) with fatty acids (three molecules).

Van Der Waals forces Also called dispersion forces, these occur when small, temporary dipoles are formed because of the random motion of electrons.

VHEPR Valence shell electron pair repulsion theory.

APPENDIX

C Additional Resources

Study Guides

Eubanks, Lucy, T., and I. Dwaine Eubanks. *Preparing for Your ACS Examination in General Chemistry: The Official Guide.* (American Chemical Society Examinations Institute, 1999).

Flowers, James L., and Theodore Silver. *Cracking the MCAT: A Thorough Review of All the MCAT Science You Need to Know to Score Higher*. (New York: The Princeton Review, 2004).

Schaum's Outline Series, *Chemistry.*

Textbooks

American Chemical Society Staff. *Chemistry in the Community (Student Edition): Chemcom, 4th Ed.* (New York: W. H. Freeman, 2000).

Bettelheim, Frederick, Jerry March, and William H. Brown. *Introduction to General, Organic, and Biochemistry, 7th Ed.* (New York: Thomson Learning, 2003).

Caret, Robert L., Katherine Denniston, and Joseph Topping. *Foundations of Inorganic, Organic and Biological Chemistry* (Dubuque, IA: William C. Brown, 1995).

Hummel, Thomas J., Steven S. Zumdahl, and Susan Arena Zumdahl. *Chemistry (Student Solutions Manual), 6th Ed.* (Houghton Mifflin, 2004).

Stoker, H. Stephen, and Barbara G. Walker. *Fundamentals of Chemistry: General, Organic and Biological, 2nd Ed.* (Englewood Cliffs, NJ: Prentice Hall, 1990).

Timberlake, Karen C. *Chemistry: An Introduction to General, Organic and Biological Chemistry, 6th Ed.* (Redwood City, CA: Benjamin-Cummings, 1995).

Zumdahl, Steven S. *Chemical Principles, 5th Ed.* (Houghton Mifflin, 2004).

ADDITIONAL ONLINE PRACTICE

Whether you need help building basic skills or preparing for an exam, visit the LearningExpress Practice Center! On this site, you can access additional practice materials. Using the code below, you'll be able to log in and take an additional chemistry practice exam. This online practice will also provide you with:

- **Immediate scoring**
- **Detailed answer explanations**
- **Personalized recommendations for further practice and study**

Log in to the LearningExpress Practice Center by using this URL: **www.learnatest.com/practice**

This is your access code: **7991**

Follow the steps online to redeem your access code. After you've used your access code to register with the site, you will be prompted to create a username and password. For easy reference, record them here:

Username: ______________________ **Password:**________________________

With your username and password, you can log in and answer these practice questions as many times as you like. If you have any questions or problems, please contact LearningExpress customer service at 1-800-295-9556 ext. 2, or e-mail us at **customerservice@learningexpressllc.com**.

NOTES

NOTES

NOTES

NOTES

NOTES

NOTES

NOTES